PHILIP'S MONTH-BY-MONTH

STARGAZING 2016

THE GUIDE TO THE NORTHERN NIGHT SKY

www.philipsastronomy.com
www.philips-maps.co.uk

HEATHER COUPER and NIGEL HENBEST are internationally recognized writers and broadcasters on astronomy, space and science. They have written more than 40 books and over 1000 articles, and are the founders of an independent TV production company specializing in factual and scientific programming.

Heather is a past President of both the British Astronomical Association and the Society for Popular Astronomy. She is a Fellow of the Royal Astronomical Society, a Fellow of the Institute of Physics and a former Millennium Commissioner, for which she was awarded the CBE in 2007. Nigel has been Astronomy Consultant to *New Scientist* magazine, Editor of the *Journal of the British Astronomical Association* and Media Consultant to the Royal Greenwich Observatory.

Heather Couper

Nigel Henbest

Published in Great Britain in 2015 by Philip's,
a division of Octopus Publishing Group Limited
(www.octopusbooks.co.uk)
Carmelite House, 50 Victoria Embankment,
London EC4Y 0DZ
An Hachette UK Company (www.hachette.co.uk)

TEXT
Heather Couper and Nigel Henbest (pages 4–53)
Robin Scagell (pages 61–64)
Philip's (pages 1–3, 54–60)

ISBN 978–1–84907–391–2

A CIP catalogue record for this book is available from the British Library.

Printed in China

Title page: *The Eagle Nebula*
(Ed Cloutman/Galaxy)

ACKNOWLEDGEMENTS

All star maps by Wil Tirion/Philip's, with extra annotation by Philip's.
Artworks © Philip's.

All photographs courtesy of Galaxy Picture Library:
Ed Cloutman *1, 32, 36;*
Paul Coleman *16;*
Jamie Cooper *24;*
Nick Hart *52;*
Ian King *44;*
Martin Lewis *28;*
Ian Papworth *62;*
Damian Peach *12, 61;*
Robin Scagell *63, 64;*
Peter Shah *8, 48;*
Ian Sharp *20;*
Dave Tyler *40.*

CONTENTS

Welcome to the world of stargazing! Within these pages, you'll find your complete guide to everything that's happening in the night sky throughout 2016 – whether you're a beginner or a seasoned astronomer.

With the 12 monthly star charts, you can find your way around the sky on any night in the year. Impress your friends by identifying celestial sights ranging from the brightest planets to the most obscure constellations! And for both the novice and the expert, we've included updates on everything that's new in 2016, from comets to shooting stars to eclipses.

THE MONTHLY CHARTS

A reliable map is just as essential for exploring the heavens as it is for visiting a foreign country. For each month, we provide two **star charts**, showing the view looking north and south. To keep the maps uncluttered, we've plotted about 200 of the brighter stars (down to third magnitude), which means you can pick out the main star patterns – the constellations. (If we'd shown every star visible on a really dark night, there'd be around 3000 stars on the charts!) We also show the ecliptic: the apparent path of the Sun in the sky, which is closely followed by the Moon and planets as well. You can use these charts throughout most of Europe, North America and northern Asia – between 40° and 60° north – though our detailed notes apply specifically to the UK and Ireland.

USING THE STAR CHARTS

It's pretty easy to use the charts. Start by working out your compass points. South is where the Sun is highest in the sky during the day; east is roughly where the Sun rises, and west where it sets. You can find north by locating the Pole Star (Polaris) at night by using the stars of the Plough (see February).

The left-hand chart then shows your view to the north. Most of the stars here are visible all year: these circumpolar constellations wheel around Polaris as the seasons progress.

Your view to the south appears in the right-hand chart; it changes much more as the Earth orbits the Sun. Leo's prominent 'Sickle' is high in the spring skies. Summer is dominated by the bright trio of Vega, Deneb and Altair. Autumn's familiar marker is the Square of Pegasus; while the stars of Orion rule the winter sky.

During the night, our perspective on the sky also alters as the Earth spins around, making the stars and planets appear to rise in the east and set in the west. The charts depict the sky in the late evening (the exact times are noted in the captions). As a rule of thumb, if you are observing two hours later, then the following month's map will be a better guide to the stars on view – though beware: the Moon and planets won't be in the right place!

THE MOON, PLANETS AND SPECIAL EVENTS

Our charts also highlight the **planets** above the horizon in the late evening. However, you may be able to spot other planets on the same night, which have either set by the time shown on the chart, or haven't risen yet. Turn to our monthly **Planets on View** notes, which describe what can be seen throughout the hours of darkness.

The position of the Full Moon is plotted each month, and also the **Moon's position** at three-day intervals before and after. The adjacent table has detailed information on the **Moon's phases**. If there's a **meteor shower** in the month, we mark the radiant – the position from which the meteors appear to stream – and describe it more fully in the **Special Events** section. Here you'll also find information on close pairings of the planets, times of the equinoxes and solstices, and – most exciting – **eclipses** of the Moon and Sun.

In addition, we've indicated the track of any **comets** known at the time of writing; though we're afraid we can't guide you to a comet that's found after the book has been printed!

Each month, we examine one particularly interesting **object**: a planet perhaps, or a star or a galaxy. We also feature a spectacular **picture** – taken by a backyard amateur – and describe how the image was captured. And we explore a fascinating **topic**, ranging from weird moons to the Christmas star.

There's a full annual overview of events in the **Solar System Almanac** on pages 54–57, along with diagrams explaining the motion of the planets and the celestial choreography that leads to an eclipse. More details about the magnitude scale, light years and the separation of objects in the sky are also given.

GETTING IN DEEPER

For each month, there's a practical **viewing tip** to help you explore the sky with the naked eye, binoculars or a telescope. If you're after 'faint fuzzies' that are too dim to appear on the charts, turn to the list of **recommended deep-sky objects** (pages 58–60). The adjacent table of 'limiting magnitude' indicates which objects are visible with your equipment.

Finally, for a round-up of what's new in observing technology go to pages 61–64, where equipment expert Robin Scagell offers advice on the planetary imaging revolution.

New for 2016, we're providing some ideas for what you can do on cloudy nights, or during the day. Follow the links in our **Citizen Science** boxes – scattered throughout the book – to join in projects at the cutting edge of astronomical research, from analyzing Martian clouds and classifying galaxies to hunting down the first radio messages from aliens....

Happy stargazing!

In a great year for stargazers, 2016 starts off with a double whammy – the brilliant stars of winter, adorning the glittering star patterns of **Orion**, **Taurus**, **Auriga**, **Gemini** and **Canis Major**, along with a comet to add to the New Year sparkle. And there's a whole smorgasbord of planetary activity in the early morning skies.

▼ *The sky at 10 pm in mid-January, with Moon positions at three-day intervals either side of Full Moon. The star positions are also correct for 11 pm at*

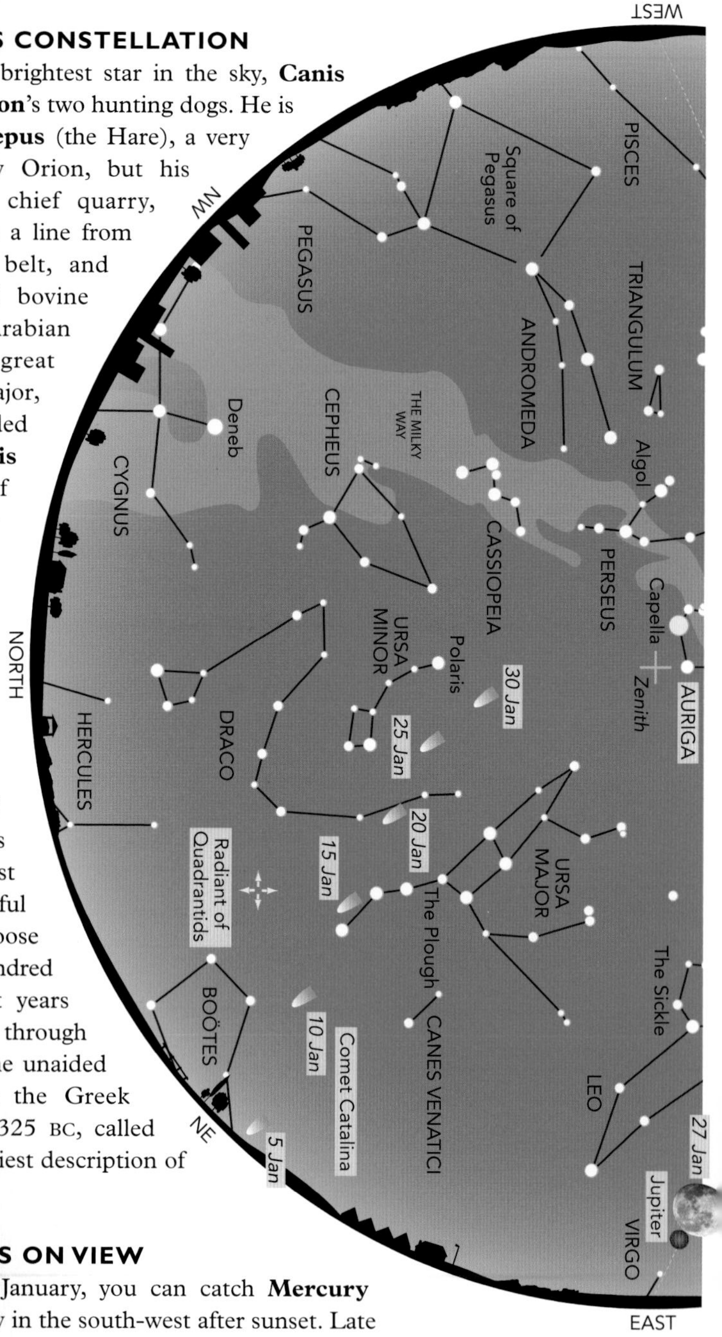

JANUARY'S CONSTELLATION

Crowned by **Sirius**, the brightest star in the sky, **Canis Major** is the larger of **Orion**'s two hunting dogs. He is represented as chasing **Lepus** (the Hare), a very faint constellation below Orion, but his main target is Orion's chief quarry, **Taurus** (the Bull) – take a line from Sirius through Orion's belt, and you'll spot the celestial bovine on the other side. Arabian astronomers accorded great importance to Canis Major, while the Indians regarded both cosmic dogs (**Canis Minor** lies to the left of Orion) as being 'watch-dogs of the Milky Way' – which runs between the two constellations.

To the right of Sirius is the star **Mirzam**. Its Arabic name means 'the announcer', because the presence of Mirzam heralded the appearance of Sirius, one of the heaven's most venerated stars. Just below Sirius is a beautiful star cluster, **M41**. This loose agglomeration of over a hundred young stars – 2300 light years away – is easily visible through binoculars, and even to the unaided eye. It's rumoured that the Greek philosopher Aristotle, in 325 BC, called it 'a cloudy spot' – the earliest description of a deep-sky object.

PLANETS ON VIEW

During the first days of January, you can catch **Mercury** (magnitude −0.1) very low in the south-west after sunset. Late

the beginning of January, and 9 pm at the end of the month. The planets move slightly relative to the stars during the month.

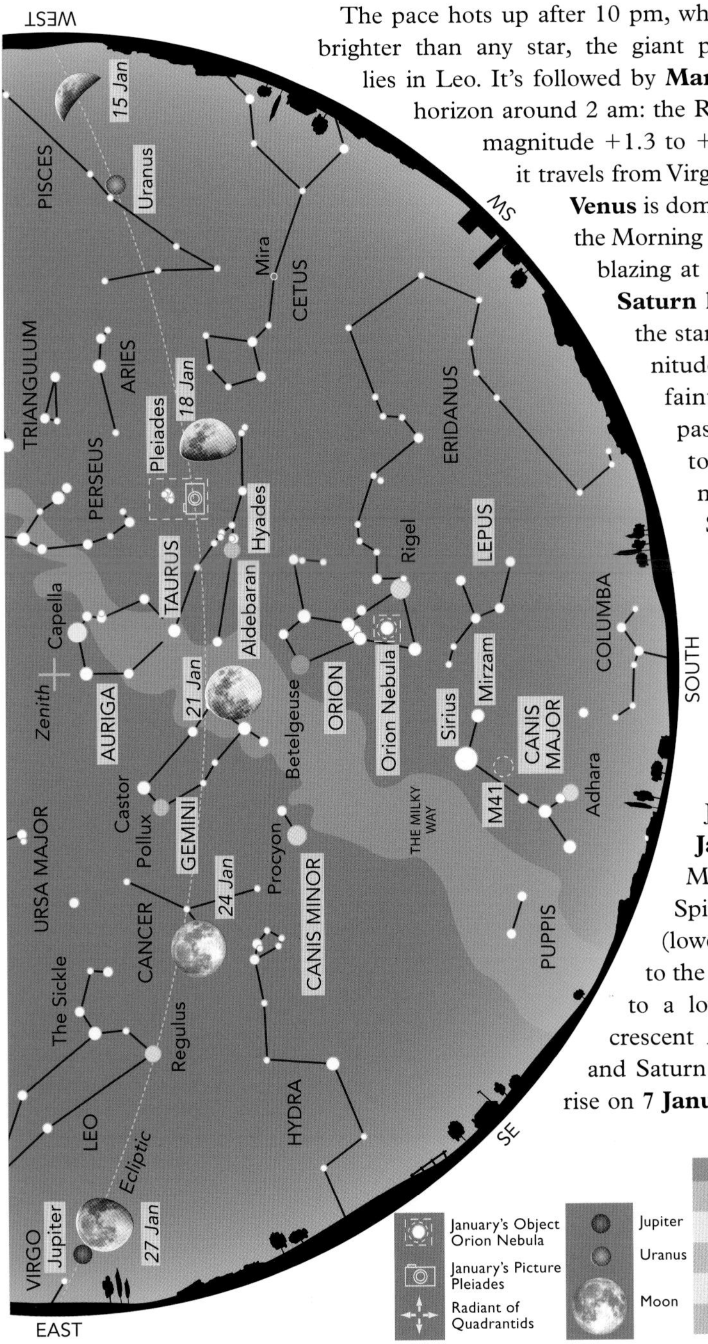

in the month, it swings round into the morning sky to join Venus (see below).

Otherwise the early evening sky contains only the dim planet **Neptune** (magnitude +7.9), which lies in Aquarius and sets around 8 pm, and the slightly brighter **Uranus** – setting about midnight – which you'll find in Pisces at magnitude +5.8.

The pace hots up after 10 pm, when **Jupiter** rises. Shining brighter than any star, the giant planet (magnitude −2.1) lies in Leo. It's followed by **Mars**, which pops above the horizon around 2 am: the Red Planet brightens from magnitude +1.3 to +0.8 during the month as it travels from Virgo into Libra.

Venus is dominating the pre-dawn sky: the Morning Star rises around 5.30 am, blazing at magnitude −3.9. Distant **Saturn** lies to the left of Venus at the start of January; with a magnitude of +0.7, it's 70 times fainter. The Morning Star passes extraordinarily close to the ringworld on the morning of **9 January** (see Special Events). Towards the close of the month. Venus is closing in on Mercury.

MOON

In the early morning of **New Year's Day**, the Moon lies near Jupiter. Before dawn on **3 January**, the Last Quarter Moon forms a triangle with Spica (lower right) and Mars (lower left); on **4 January**, it's to the left of Mars. We're treated to a lovely view of the slender crescent Moon with Venus (right) and Saturn (lower right) before sunrise on **7 January**. On **19 January**, the

MOON		
Date	Time	Phase
2	5.30 am	Last Quarter
10	1.30 am	New Moon
16	11.26 pm	First Quarter
24	1.46 am	Full Moon

Moon occults several members of the Hyades star cluster, plus Aldebaran (see Special Events). The star near the Moon on **25 January** is Regulus. Jupiter and the waning Moon make a striking pair on **27 January**.

SPECIAL EVENTS

As you head home after a New Year's Eve party, you may notice the bright star Arcturus looking distinctly fuzzy. Don't worry: there's nothing wrong with your eyesight! **Comet Catalina** is passing almost directly in the front of Arcturus that night. Discovered by the Catalina Sky Survey in 2013, this comet zoomed by the Sun late last year, and is now heading back to deep space. Starting January at magnitude +4, Comet Catalina fades during the month to magnitude +5, as it speeds towards Polaris, the Pole Star.

On **2 January**, at 10.49 pm, the Earth is at perihelion, its closest point to the Sun – a 'mere' 147 million kilometres away.

The night of **3/4 January** sees the maximum of the **Quadrantid** meteor shower, tiny particles of dust shed by the old comet 2003 EH_1 that burn up – often in a blue or yellow streak – as they enter the Earth's atmosphere. Look out for them around midnight, before the Moon rises.

Before dawn on **9 January**, brilliant Venus passes extremely close to Saturn: just 5 arc minutes away. With the naked eye, you may not even notice Saturn so near to a planet that's almost a hundred times brighter. Binoculars will show the conjunction well; through a small telescope, you can see the globe of Venus and Saturn's rings in the same field of view.

Watch the Moon carefully – preferably with binoculars or a small telescope – on **19 January**, as it moves right in front of the **Hyades**. Depending on your location, you'll see the Moon either occult – or narrowly miss – several of the cluster's brighter stars, and also the neighbouring red giant **Aldebaran**.

▼ *Peter Shah captured the Pleiades using an Orion Optics AG8 200 mm f/3.8 Advanced Newtonian Astrograph telescope and a Starlight Xpress SXVF-H16, with Astronomik type 1 RGB filters. Peter's exposures were 600 seconds per subframe; 60 minutes in each filter. The total exposure time was 3 hours.*

JANUARY'S OBJECT

Below **Orion**'s Belt lies a fuzzy patch – easily visible to the unaided eye in dark skies. Through binoculars, or a small telescope, the patch looks like a small cloud in space. It is a cloud – but at 24 light years across, it's hardly petite. Only the distance of the **Orion Nebula** – 1300 light years – diminishes it. It's part of a vast region of starbirth in Orion, and the nearest region to Earth where heavyweight stars are being born. This 'star factory' contains at least

700 fledgling stars, which have just hatched out of immense dark clouds of dust and gas: most are visible only with a telescope that picks up infrared (heat) radiation.

JANUARY'S PICTURE

The small but perfectly formed **Pleiades** star cluster is as much a feature of our winter skies as the magnificent constellation of Orion. The 'Seven Sisters' visible to the naked eye comprise just a fraction of the 1000 stars making up the cluster (see December's Object). They are a delightful sight in binoculars, and magnificent when you view them through a wide-field telescope.

As this image shows, the stars are hot and blue – fledglings on the celestial age scale. Long exposures also reveal glorious nebulosity around the stars. The cluster is passing through a thick region of interstellar dust, which reflects the stars' brilliant blue light.

Viewing tip

When you first go out to observe, you may be disappointed at how few stars you can see in the sky. But wait for around 20 minutes and you'll be amazed at how your night vision improves. One reason for this 'dark adaption' is that the pupil of your eye gets larger to make the best of the darkness. More importantly, in dark conditions the retina of your eye builds up much bigger reserves of rhodopsin, the chemical that responds to light.

JANUARY'S TOPIC
Spotty planets

Mighty Jupiter dominates the late night sky – the view of the stripy, tangerine-shaped globe of our largest planet through a telescope is unforgettable. But strange things have been happening to the giant world in recent years. Its famous 'Great Red Spot' – a vast storm swirling in the clouds of the planet's southern hemisphere – has been shrinking.

But astronomers don't think that it will fade away forever. It's still bigger than the Earth; it is powered by winds that rage at over 430 km/h; and its red colour is believed to be due to atmospheric gases blasted by the Sun's ultraviolet light.

Gas giant worlds like Jupiter are renowned for their spots. Jupiter's neighbour – ringworld Saturn – erupted spectacularly in 1933. The famous stage and screen comedian Will Hay – also an amateur astronomer – discovered a huge white spot on the planet, which has since faded.

The next world out, Uranus, has been criticized for its blandness. When the Voyager 2 space probe swept past the planet in 1986, it revealed a gaseous grey-green globe devoid of any features. But things are hotting up! As its seasons change, streaks and clouds have started to appear in its atmosphere. In August 2014, giant storms erupted on the planet.

The outermost planet, Neptune, put on a splendid show for Voyager 2's cameras in 1989. Much more active than its twin Uranus, it sported a giant dark spot at its equator, with winds blowing at speeds of 2000 km/h – the fastest in the Solar System.

The first signs of spring are now on the way. The winter star patterns are drifting towards the west, as a result of our annual orbit around the Sun. Imagine: you're on a fairground carousel, circling round on your horse, and looking out around you. At times you spot the ghost train; sometimes you see the roller-coaster; and then you swing past the candy-floss stall. So it is with the sky: as we circle our local star, we get to see different stars and constellations with the changing seasons.

▼ *The sky at 10 pm in mid-February, with Moon positions at three-day intervals either side of Full Moon. The star positions are also correct for*

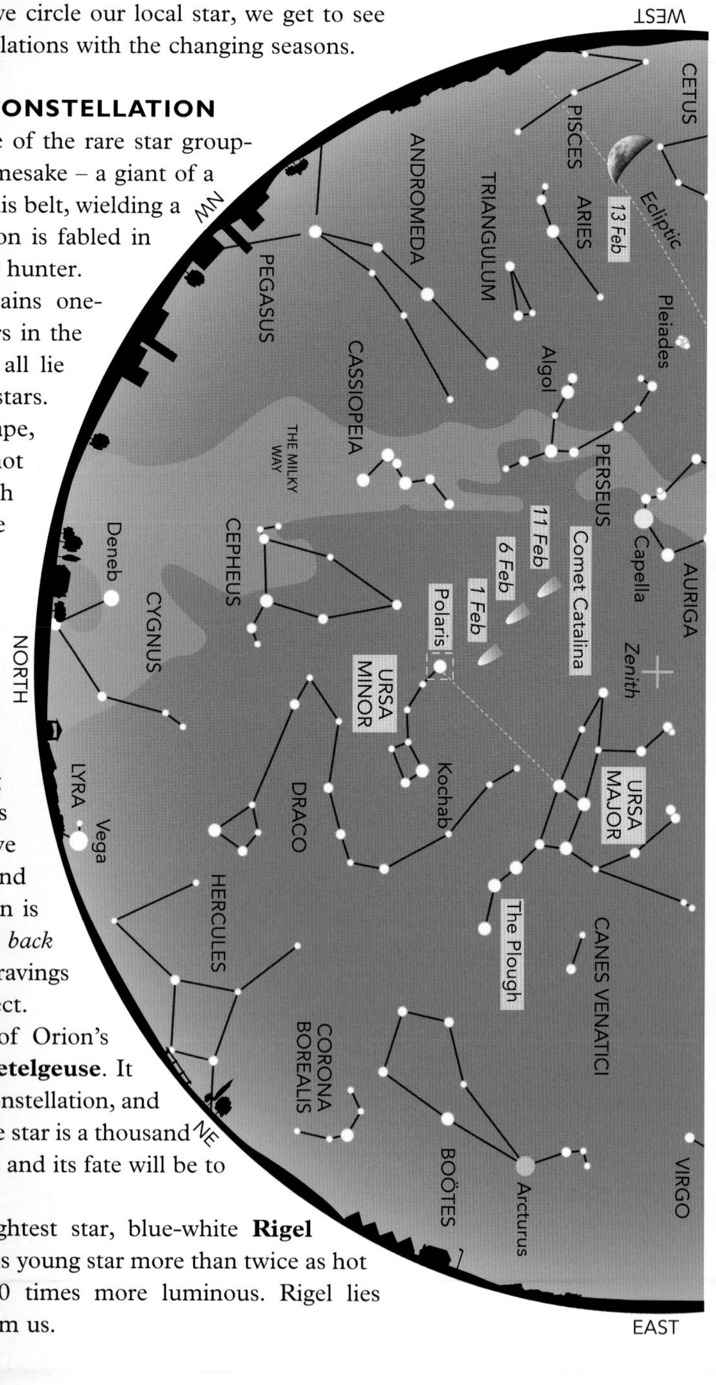

FEBRUARY'S CONSTELLATION

Spectacular **Orion** is one of the rare star groupings that looks like its namesake – a giant of a man with a sword below his belt, wielding a club above his head. Orion is fabled in mythology as the ultimate hunter.

The constellation contains one-tenth of the brightest stars in the sky: its seven main stars all lie in the 'top 70' brilliant stars. Despite its distinctive shape, most of these stars are not closely associated with each other – they simply line up, one behind the other.

Closest is the star that forms the hunter's right shoulder, **Bellatrix**, at 250 light years. And an interesting puzzle here! One of our readers pointed out that if Orion is facing us, Bellatrix should be his *left* shoulder. But we've scoured the evidence, and have discovered that Orion is as often depicted with his *back* to us in mythological engravings – so 'right' would be correct.

Next in the hierachy of Orion's superstars is blood-red **Betelgeuse**. It lies at the top left of the constellation, and is 640 light years away. The star is a thousand times larger than our Sun, and its fate will be to explode as a supernova.

The constellation's brightest star, blue-white **Rigel** (Orion's foot), is a vigorous young star more than twice as hot as our Sun, and 125,000 times more luminous. Rigel lies around 860 light years from us.

11 pm at the beginning of February, and 9 pm at the end of the month. The planets move slightly relative to the stars during the month.

Saiph, which marks the constellation's other foot, is around 650 light years distant. The two outer stars of the belt, **Alnitak** (left) and **Mintaka** (right), lie 700 and 690 light years away, respectively.

We travel 1300 light years from home to reach the middle star of the belt, **Alnilam**. And at the same distance, we see the stars of the 'sword' hanging below the belt – the lair of the great **Orion Nebula** – an enormous star-forming region 24 light years across (see January's Object and December's Picture).

PLANETS ON VIEW

Brilliant **Jupiter** is now lording it over the evening skies, rising about 7.30 pm and lying under the hindquarters of the celestial lion, Leo. At magnitude −2.3, the giant planet dominates the evening sky.

At the opposite extreme is distant **Uranus**, hardly visible to the naked eye. The seventh planet (magnitude +5.9) skulks in Pisces and sets around 10 pm.

After midnight, the planetary scene livens up. **Mars** rises at 1 am in the south-east, shining at magnitude +0.5 in Libra. **Saturn** follows the Red Planet above the horizon around 3 am. Lying in Ophiuchus, the ringed planet is currently a near-twin to Mars in brightness (magnitude +0.6).

Both are well out-classed when brilliant **Venus** rises, around 6 am. Blazing at magnitude −3.8, the Morning Star is slipping down into the dawn twilight and is difficult to spot by the end of the month.

February's Object
Polaris

February's Picture
Jupiter

MOON		
Date	Time	Phase
1	3.28 am	Last Quarter
8	2.39 pm	New Moon
15	7.46 am	First Quarter
22	6.20 pm	Full Moon

◀ *Damian Peach photographed the giant planet Jupiter from Barbados, on 26 February 2015. The image was taken on a C14 (355 mm) Celestron Schmidt-Cassegrain reflector using a Barlow lens to increase the focal length to f/25. The camera was a ZWO ASI174MM mono webcam, with RGB filters.*

The 'star' to the left of Venus is its smaller and fainter sibling, **Mercury** (magnitude +0.1), which reaches maximum western elongation on **7 February**.

Neptune is too close to the Sun to be visible this month.

MOON

Mars lies below the Last Quarter Moon on the morning of **1 February**. You'll find the crescent Moon near Saturn before dawn on **3 and 4 February**. The crescent Moon meets Venus and Mercury early on **6 February** (see Special Events). The brilliant object near the Moon on **23 February** is Jupiter. The Moon passes above Spica in the early hours of **27 February**.

SPECIAL EVENTS

During the first few nights of February, use binoculars to catch the fading **Comet Catalina** (see January's Special Events) at around magnitude +5 near Polaris; by the end of the month, you'll need a telescope to spot it.

There's a lovely sight in the south-east, just before dawn on **6 February**, when the narrowest crescent Moon hangs above brilliant Venus, with Mercury to the lower left.

FEBRUARY'S OBJECT

The Pole Star – **Polaris** – is a surprisingly shy animal, coming in at the modest magnitude of +2.1. You can find it by following the two end stars of the **Plough** (see chart) in

Citizen Science: Martian weather

Explore the frozen southern polar cap on Mars – in images largely unseen by human eyes before – and make your own contribution to unravelling the Red Planet's enigmatic weather. In the Planet Four project, you peruse the most detailed views from the Mars Reconnaissance Orbiter, on the trail of tantalizing dark wedges and blobs on the surface, probably erupted by dark geysers and blown sideways by the wind – along with mysterious spider-shaped systems of cracks.

http://www.planetfour.org/

Viewing tip
If you want to stargaze at this most glorious time of year, dress up warmly! Lots of layers are better than just a heavy coat, as they trap more air close to your skin, while heavy-soled boots with two pairs of socks stop the frost creeping up your legs. It may sound anorakish, but a woolly hat prevents a lot of your body heat escaping through the top of your head. And – alas – no hipflask of whisky. Alcohol constricts the veins and makes you feel even colder.

Ursa Major (the Great Bear). Polaris lies at the end of the tail of the Lesser Bear (**Ursa Minor**), and it pulsates in size, making its brightness vary slightly over a period of four days. But its importance throughout recent history centres on the fact that Earth's north pole points towards Polaris, so we spin 'underneath' it. It remains almost stationary in the sky, and acts as a fixed point for both astronomy and navigation. Over a 26,000-year period, the Earth's axis swings around like an old-fashioned spinning top – a phenomenon called 'precession' – so our 'pole stars' change with time.

FEBRUARY'S PICTURE

This stunning portrait of the largest world in our Solar System – **Jupiter** – was taken by a British astronomer, Damian Peach, observing from Barbados, where sky conditions are excellent. The huge gas giant spins so rapidly that its clouds are drawn out into streaks by the speed of the planet's rotation. Jupiter is a delight to observe: its atmosphere is ever-changing. And that's also true of Jupiter's Great Red Spot (centre). A constant phenomenon of the planet for at least three centuries, the spot has been shrinking over the past few years. Once three times the size of the Earth, it now measures in at roughly the diameter of our planet. Will it disappear? Astronomers are confident that it won't – but instead, continue in its slimmed-down mode.

FEBRUARY'S TOPIC
Leap year

This year, February is blessed with an extra day (as anyone born on **29 February** is all too aware!), making 2016 a 'leap year'. We have leap years because there aren't an exact number of days in a year: in fact, a year is 365.2422 days long. Julius Caesar declared that every fourth year should have an extra day (added to poor February, as it's the shortest month). That makes an average of 365.25 days per year – which still isn't quite right. In 1582, Pope Gregory XIII changed the rules, so that a century year is a leap year only if you can divide it by 400. So 2000 was a leap year, while 2100 won't be. Over many centuries, the 'Gregorian calendar year' averages out to 365.2425 days. That's pretty close to the actual length of the year. But, amazingly, back in AD 1079, the Persian astronomer and poet Omar Khayyam devised a calendar with a pattern of eight leap years spread over a 33-year period, and this calendar – still used in Iran – averages to 365.2424 days per year, which is more accurate than our Gregorian calendar!

Spring is here! On **20 March**, we celebrate the Vernal Equinox, when day becomes longer than the night; while British Summer Time starts on **27 March**. Though the nights may be shorter, there's plenty going on in the heavens. Check out brilliant Jupiter and – later in the night – Mars and Saturn.

▼ The sky at 10 pm in mid-March, with Moon positions at three-day intervals either side of Full Moon. The star positions are also correct for 11 pm at

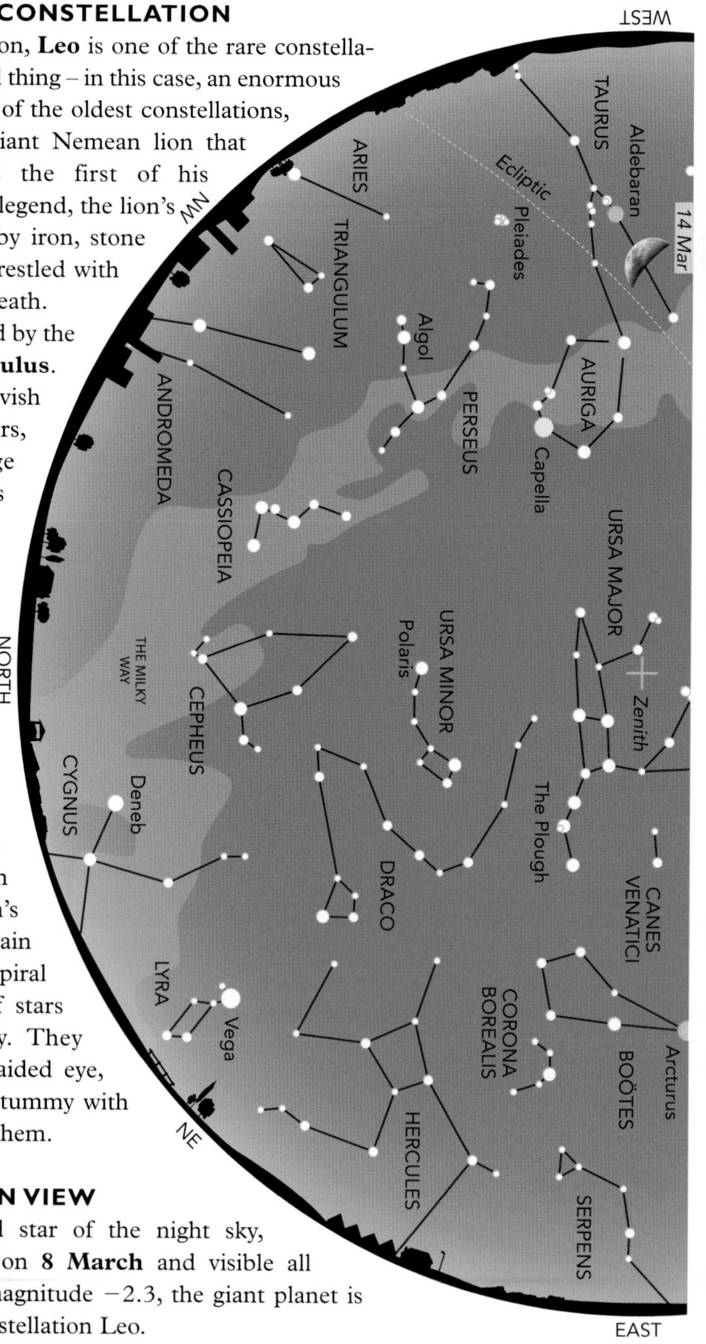

MARCH'S CONSTELLATION

Like the fabled hunter Orion, **Leo** is one of the rare constellations that resembles the real thing – in this case, an enormous crouching lion. Leo is one of the oldest constellations, and commemorates the giant Nemean lion that Hercules slaughtered as the first of his '12 labours'. According to legend, the lion's flesh couldn't be pierced by iron, stone or bronze – so Hercules wrestled with the lion and choked it to death.

The lion's heart is marked by the first-magnitude star **Regulus**. This celestial whirling dervish spins around in just 16 hours, making its equator bulge remarkably. Rising upwards is '**the Sickle**', a back-to-front question mark that delineates the front quarters, neck and head of Leo. A small telescope shows that **Algieba**, the star which makes up the lion's shoulder, is actually a beautiful close double star.

The other end of Leo is home to **Denebola**, which means in Arabic 'the lion's tail'. Just underneath the main 'body' of Leo are several spiral galaxies – nearby cities of stars like our own Milky Way. They can't be seen with the unaided eye, but sweep along the lion's tummy with a small telescope to reveal them.

PLANETS ON VIEW

Jupiter is the undisputed star of the night sky, at opposition to the Sun on **8 March** and visible all night long. At a dazzling magnitude −2.3, the giant planet is strutting its stuff in the constellation Leo.

the beginning of March, and 10 pm at the end of the month (after BST begins). The planets move slightly relative to the stars during the month.

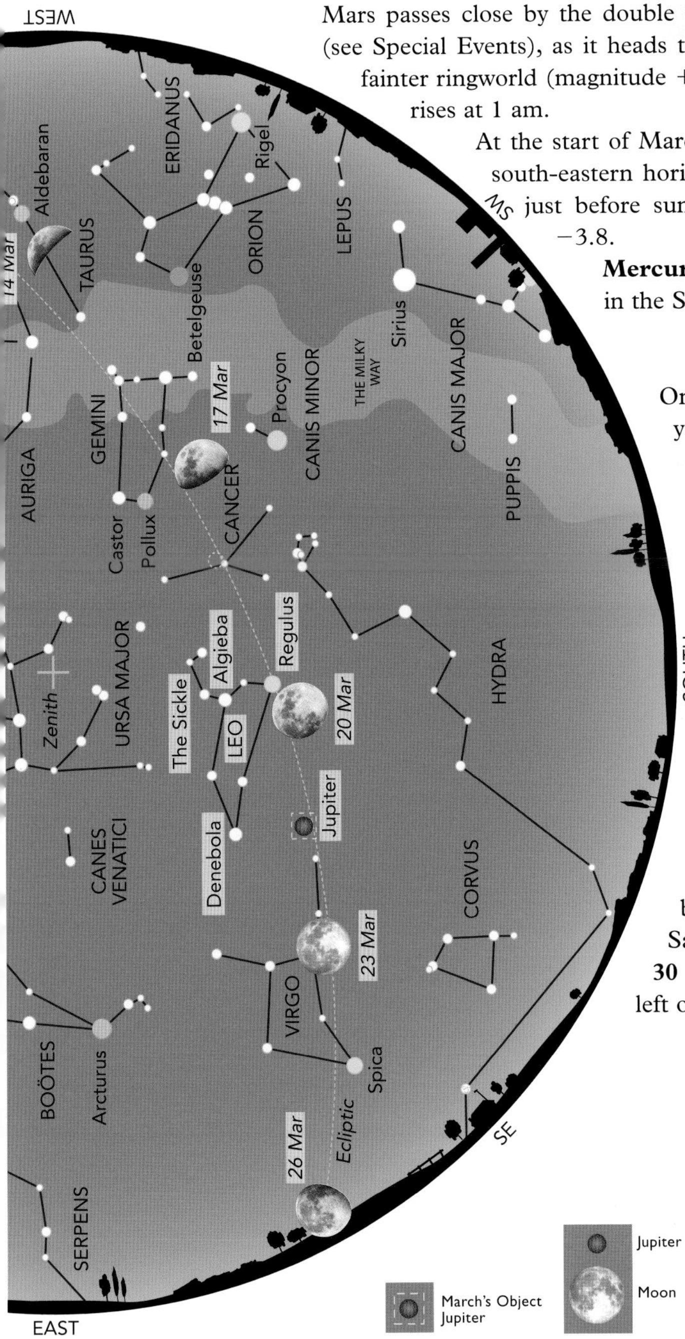

Dim **Uranus** (magnitude +5.9) lies in Pisces; setting around 8 pm, the distant world disappears into the evening twilight by the end of March.

Mars is rising in the south-east just after midnight. During the month, the Red Planet doubles in brightness, from magnitude +0.3 to −0.5, and travels from Libra into Scorpius. Mars passes close by the double star Graffias on **16 March** (see Special Events), as it heads towards **Saturn**. The rather fainter ringworld (magnitude +0.5) lies in Ophiuchus and rises at 1 am.

At the start of March – and with a really clear south-eastern horizon – you may spot **Venus** just before sunrise, shining at magnitude −3.8.

Mercury and **Neptune** are hidden in the Sun's glare this month.

WEST
ERIDANUS
Rigel
Aldebaran
TAURUS
ORION
LEPUS
SW
14 Mar
Betelgeuse
Sirius
THE MILKY WAY
CANIS MAJOR
CANIS MINOR
Procyon
17 Mar
GEMINI
AURIGA
CANCER
Castor
Pollux
PUPPIS
Regulus
URSA MAJOR
Algieba
Zenith
The Sickle
LEO
20 Mar
HYDRA
SOUTH
Jupiter
Denebola
CANES VENATICI
CORVUS
23 Mar
VIRGO
BOÖTES
Arcturus
Spica
Ecliptic
26 Mar
SE
SERPENS
EAST
March's Object Jupiter
Jupiter
Moon

MOON

On the morning of **1 March**, you'll find Mars to the right of the Last Quarter Moon; and passes Saturn early on **2 March**. In the dawn twilight of **7 March**, it will be a challenge to spot the narrowest crescent Moon above Venus. The Moon lies near Regulus on **20 March**; it's next to brilliant Jupiter on **21 March**. The star below the Moon on **24 March** is Spica. In the early hours of **29 March**, the Moon lies between Mars (right) and Saturn (left). Before dawn on **30 March**, the Moon is to the left of Saturn.

MOON

Date	Time	Phase
1	11.10 pm	Last Quarter
9	1.54 am	New Moon
15	5.03 pm	First Quarter
23	12.01 pm	Full Moon
31	4.17 pm	Last Quarter

SPECIAL EVENTS

Jupiter is at opposition on **8 March** (see March's Object).

A total eclipse of the Sun on **9 March** is visible from a narrow region that stretches from Sumatra, through Borneo and Sulawesi, to end in the Pacific Ocean north of Hawaii. People in Indochina, northern Australasia and the north-west Pacific will witness a partial eclipse. Find more details at eclipse.gsfc.nasa.gov/solar.html.

There's a lovely sight through a small telescope in the early hours of **16 March**, as **Mars** passes only 9 arc minutes from the double star Graffias (each star is actually a close triple, so Graffias really consists of six stars!).

The Vernal Equinox, on **20 March** at 4.30 am, marks the beginning of spring, as the Sun moves up to shine over the northern hemisphere.

27 **March**, 1.00 am: British Summer Time starts – don't forget to put your clocks forward (the mnemonic is 'Spring forward, Fall back').

MARCH'S OBJECT

Jupiter reaches opposition on **8 March**, lying opposite the Sun, and is at its closest to the Earth. 'Close' is a relative term, though! The giant planet lies about 700 million kilometres away.

At 143,000 kilometres in diameter, Jupiter is the biggest world in our Solar System. It could contain 1300 Earths – and the cloudy gas giant is very efficient at reflecting sunlight. Jupiter now is shining at a dazzling magnitude −2.3, and it's a fantastic target for stargazers, whether you're using your unaided eyes, binoculars or a small telescope.

Despite its size, Jupiter spins faster than any other planet, in 9 hours 55 minutes. As a result, its equator bulges outwards – through a small telescope, it looks like a squashed orange crossed with an old-fashioned humbug. The stripes are cloud belts of ammonia and methane stretched out by the planet's dizzy spin.

Jupiter commands a family of almost 70 moons. The four biggest are visible in good binoculars, and even – to the really sharp-sighted – to the unaided eye. These are worlds in their own right: Ganymede is actually bigger than the planet Mercury.

▼ *This is a wide-angle view of the total solar eclipse of 29 March 2006, seen from Turkey, and taken by Paul Coleman. He took the photo on ISO 400 film using a Pentax Spotmatic camera. The exposure time was 1 second at f/5.6 with a 28 mm lens.*

Viewing tip

This is the ideal time of year to tie down your compass points – the directions of north, south, east and west as seen from your observing site. North is easy – just latch on to Polaris, the Pole Star (see February's Object). And at noon, the Sun is always in the south. But the useful extra in March is that we hit the Spring (Vernal) Equinox, when the Sun rises due east, and sets due west. So remember those positions relative to a tree or house around your horizon.

MARCH'S PICTURE

Every **total solar eclipse** is different. The Sun's atmosphere – its ghostly corona – changes in shape. It's spiky and sensational when the Sun is at its most magnetically active – and plain boring when it's sulking at sunspot minimum. That's why eclipses are addictive. You never get to see the same thing twice. We've seen six, from Indonesia to Hawaii to Aruba in the Caribbean. That's as nothing as compared to American eclipse-chaser Glenn Schneider, who's seen 32 total eclipses!

We saw this total eclipse (March 2006) from a desert in Egypt – with the whole of the Egyptian government in attendance, sleeping in makeshift marquees! The eclipse track also passed through Turkey, and we love the atmospheric image that Paul Coleman captured from there.

MARCH'S TOPIC
Total solar eclipse

Nothing prepares you for a total eclipse of the Sun. Yes: many of us have seen partial eclipses, when the disc of the Moon covers most of the Sun's disc. We then see our local star as a crescent: but it's not the real thing. To get the true experience, you have to travel to a total eclipse of the Sun, where the Moon and Sun exactly overlap.

The Sun and Moon cross paths in the sky. The Moon is 400 times smaller than the Sun – but the Sun is 400 times further away. It's an incredible astronomical coincidence. And you have to be in exactly the right place on Earth to witness the astonishing alignment.

And that's when you get to see the frightening spectacle of totality. The dark Moon covers up the Sun, and – in a flash – the Sun's pearly atmosphere emerges behind the Moon's shadow. Around the rim of the Sun, you can see its beautiful crimson inner atmosphere: the chromosphere.

The whole spectacle is awesome: the eclipsed Sun looks like a Chinese dragon mask. Where has the Sun gone? When will it return? No wonder eclipses caused panic in the past.

But soon – within minutes – it's over. A chink of sunlight emerges from behind a crater on the Moon, creating the glorious 'diamond ring effect'. And then it's over – until your next total eclipse!

If you're a hooked eclipse-chaser, it's going to mean a trip to Indonesia on **9 March**. There, the eclipse will be total for over three minutes. Time to get in touch with your travel agent!

It's a great month for the planets, with a chance to spot most of our fellow worlds from elusive Mercury – setting soon after the Sun – out to distant Saturn, with giant Jupiter reigning supreme in the evening sky. This brilliant planet lies within the celestial lion, **Leo**, and next to its fellow spring constellation **Virgo** – which looks just like a giant letter 'Y' in the heavens!

▼ *The sky at 11 pm in mid-April, with Moon positions at three-day intervals either side of Full Moon. The star positions are also correct for midnight at the beginning of*

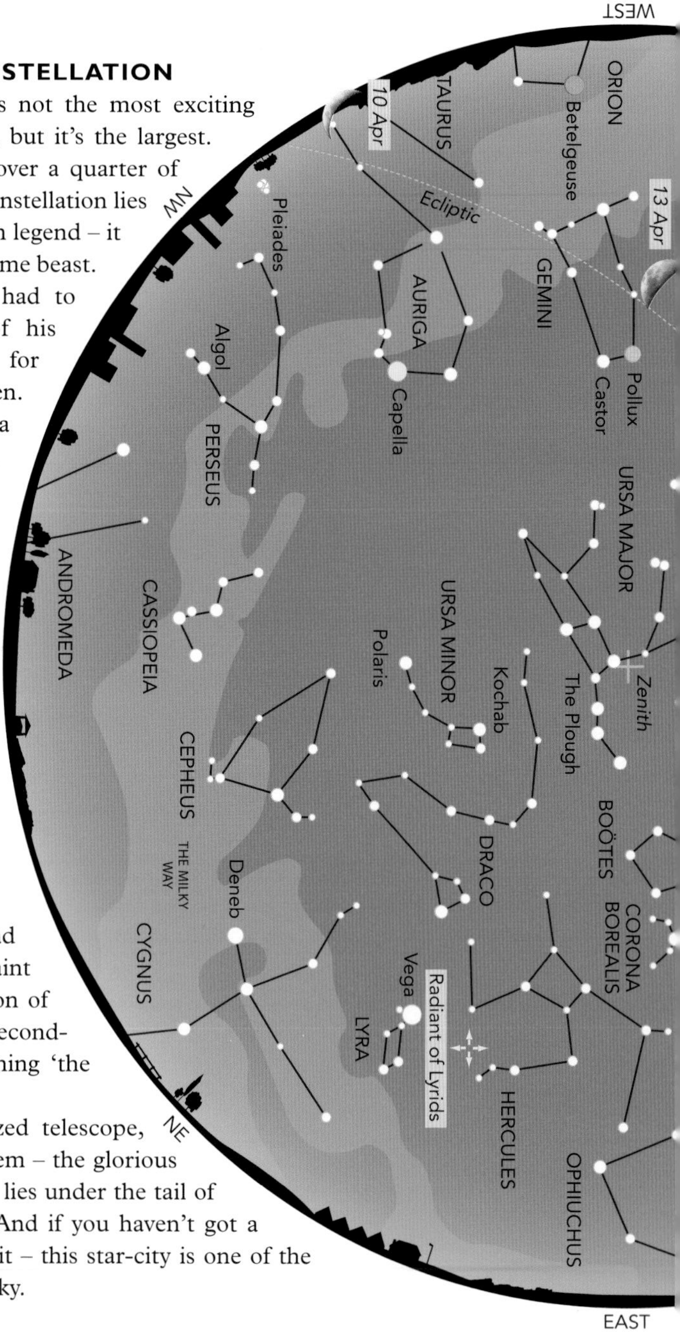

APRIL'S CONSTELLATION

Hydra (the Water Snake) is not the most exciting constellation in the heavens, but it's the largest. Hydra's faint stars straggle over a quarter of the sky (100 degrees). The constellation lies south of mighty **Leo**, and – in legend – it is meant to represent a fearsome beast.

The superhero Hercules had to slay the Hydra as one of his '12 labours' – a penance for killing his wife and children. But dispatching the Hydra wasn't easy. Even grown men died of fright when they saw the monster. And it had the irksome habit of growing numerous heads – if one was chopped off, three would grow back!

Undaunted, Hercules hacked away the extra heads, cauterizing the stumps with burning branches. Finally, he severed the last immortal head and buried it, still hissing, under a stone.

In the heavens, Hydra's head is a pretty grouping of faint stars beneath the constellation of **Cancer**. Its main star is second-magnitude **Alphard** – meaning 'the solitary one'.

If you have a medium-sized telescope, search out Hydra's hidden gem – the glorious face-on spiral galaxy **M83**. It lies under the tail of the elongated Water Snake. And if you haven't got a telescope, just google-image it – this star-city is one of the most beautiful sights in the sky.

April, and 10 pm at the end of the month. The planets move slightly relative to the stars during the month.

PLANETS ON VIEW

Elusive **Mercury** is putting on its best evening appearance of the year, though you'll need a really clear horizon to the north-west to spot the little beast – binoculars are an enormous help. You may see the innermost planet as early as **8 April**, to the right of the narrowest sliver of a crescent Moon. After that, Mercury is visible between 9 and 10 pm until a few days after its greatest eastern elongation on **18 March**. During this period, the planet fades dramatically, from magnitude −1.0 to +1.0.

Jupiter reigns supreme over the evening skies, from its perch under the main stars of Leo. Magnificent at magnitude −2.2, the giant planet is setting around 5 am.

After midnight, you can't miss **Mars** in the south-east, on the borders of Scorpius and Ophiuchus. During April, the stunning Red Planet brightens from magnitude −0.5 to −1.5 as the Earth speeds towards it. Rising half an hour later – and lying to the left of Mars – you'll find **Saturn**, shining a rather duller magnitude +0.4 in Ophiuchus. The star forming a triangle to the lower right of Mars and Saturn is red giant Antares.

Keen observers with telescopes can now see **Neptune** in the early morning skies: at magnitude +7.9, the distant planet is rising around 5 am in Aquarius.

Venus and **Uranus** are too close to the Sun to be visible this month.

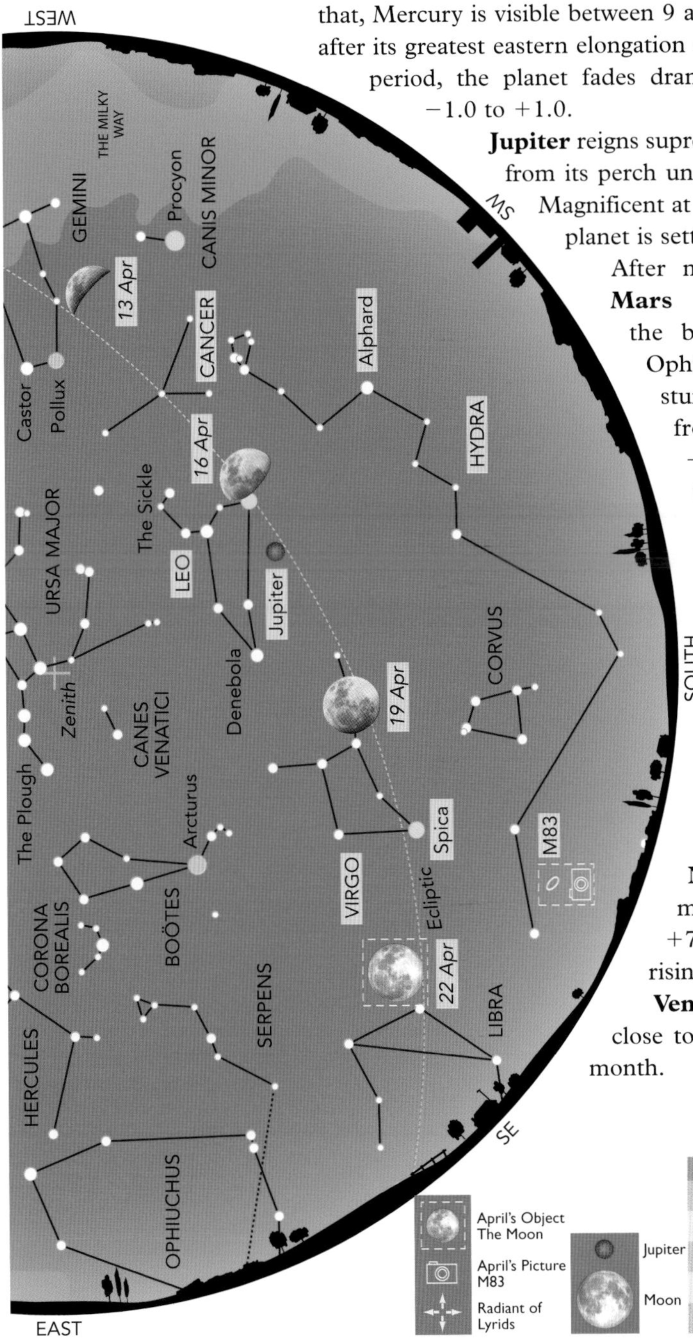

April's Object The Moon
April's Picture M83
Radiant of Lyrids
Jupiter
Moon

MOON		
Date	Time	Phase
7	12.23 pm	New Moon
14	4.59 am	First Quarter
22	6.23 am	Full Moon
30	4.28 am	Last Quarter

◀ In just under 32 hours of exposure consisting of 94 separate 20-minute exposures through R, G, B and H-alpha filters, Ian Sharp took his sensational image of M83 through the 317 mm telescope in Siding Spring, New South Wales, Australia. The Apogee F16M-D9 mono camera has a chip 36 mm square, so it is even larger than a 35 mm film format.

MOON

In the dusk twilight of **8 April**, the slimmest crescent Moon lies to the left of Mercury. On **10 April**, you'll see the Moon right in front of the Hyades when it grows dark, passing very close to Aldebaran at moonset. The star near the Moon on **16 April** is Regulus. Even more spectacularly, on **17 April** glorious Jupiter lies to the left of the Moon. In the early hours of **25 April**, you'll find the waning Moon near the planets Mars (right) and Saturn (left).

SPECIAL EVENTS

21/22 April: It's the maximum of the **Lyrid** meteor shower, which appears to emanate from the constellation of Lyra. Unfortunately, this year the whole shooting-star match is washed out by bright moonlight.

APRIL'S OBJECT

The **Moon** is our nearest celestial companion, lying a mere 384,400 kilometres away. It took the Apollo astronauts only three days to reach it! And at 3476 kilometres across, it's so large when compared to Earth that – from space – the system would look like a double planet.

But the Moon couldn't be more different from our verdant Earth. Almost bereft of an atmosphere, it has been exposed to

Citizen Science: Galaxy identification parade

Be the first person to examine a distant galaxy and decide whether it's a glorious spiral of young stars and gas, a dull elliptical ball of elderly stars, or a cosmic smash-up between galaxies. Galaxy Zoo parades in front of you a huge array of images from the Hubble Space Telescope and ground-based instruments. Early results showed that home-based participants are as reliable as trained professional astronomers: Galaxy Zoo volunteer Hanny van Arkel even found a new kind of young galaxy that has been activated by a nearby quasar.
http://www.galaxyzoo.org/

> **Viewing tip**
> It's always fun to search out the 'faint fuzzies' in the sky: galaxies, star clusters and nebulae. But don't even think about it near the time of Full Moon – its light will drown them out. The best time to observe these deep-sky objects is just before or after New Moon. When the Moon is bright, focus on planets, bright double stars – and, of course, the Moon itself. Plan your observing by checking the Moon phases timetables in the book.

bombardment by meteorites and asteroids throughout its life. Even with the unaided eye, you can see the evidence. The 'face' of the 'Man in the Moon' consists of huge craters created by asteroid hits 3.8 billion years ago.

Through binoculars or a telescope, the surface of the Moon looks amazing – as if you're flying over it. But don't observe our satellite when it's Full: the light is flat and swamps its features. It's best to roam the Moon when it's a crescent or half-lit, and see the sideways-on shadows highlighting its dramatic relief.

APRIL'S PICTURE

M83 is known as the Southern Pinwheel galaxy. Just visible in the northern hemisphere – in the constellation of Hydra – this stunning spiral star-city has hosted six supernovae in recent years. It has a central bar in its nucleus, like our own Milky Way.

Ian Sharp – from the UK – used the technique of 'remote observing' (which we explained in *Philip's Stargazing 2014*) to control a telescope in Australia that obtained this image.

APRIL'S TOPIC
Redshift

Remember the time when you last heard a police car or ambulance rushing past, siren wailing? And how the pitch of the siren sounded higher as it approached – only to drop as it passed you? The soundwaves from the siren were compressed to higher frequencies as they approached you – and then they dropped to lower notes as the vehicle sped away.

The same is true of light: the fastest waves in the Universe. And the compression in this case means that light from an object in space approaching you is shifted to the blue (high-frequency) end of the spectrum; in receding objects, their light is pushed to the red (low-frequency) end.

The 'redshift' has been invaluable in understanding the structure and evolution of the Universe. In the 1920s, American astronomer Edwin Hubble (the Hubble Space Telescope is named after him) observed distant galaxies with the biggest telescopes of his day – and concluded that they almost all showed redshifts.

There was only one conclusion: that the Universe was expanding. And with the fantastic technology we have today, scientists can pinpoint the reason for our ever-growing cosmos. By backtracking the motions of the galaxies, astronomers now agree that our Universe was born in an infernal, hot and energetic 'big bang' – nearly 14 billion years ago.

Mars is back! After two years skulking out of sight, the Red Planet makes its brilliant mark on the heavens this month, at its brightest in ten years. Mars is so dazzling that it's outshining all the stars, and even putting giant planet Jupiter to shame. Of the stars on view, we have a soft spot for orange-coloured **Arcturus**, the brightest star in the constellation of **Boötes** (the Herdsman), who shepherds the two bears – **Ursa Major** and **Ursa Minor** – through the heavens. Its appearance is sure sign that summer is on the way!

▼ The sky at 11 pm in mid-May, with Moon positions at three-day intervals either side of Full Moon. The star positions are also correct for midnight at the beginning of

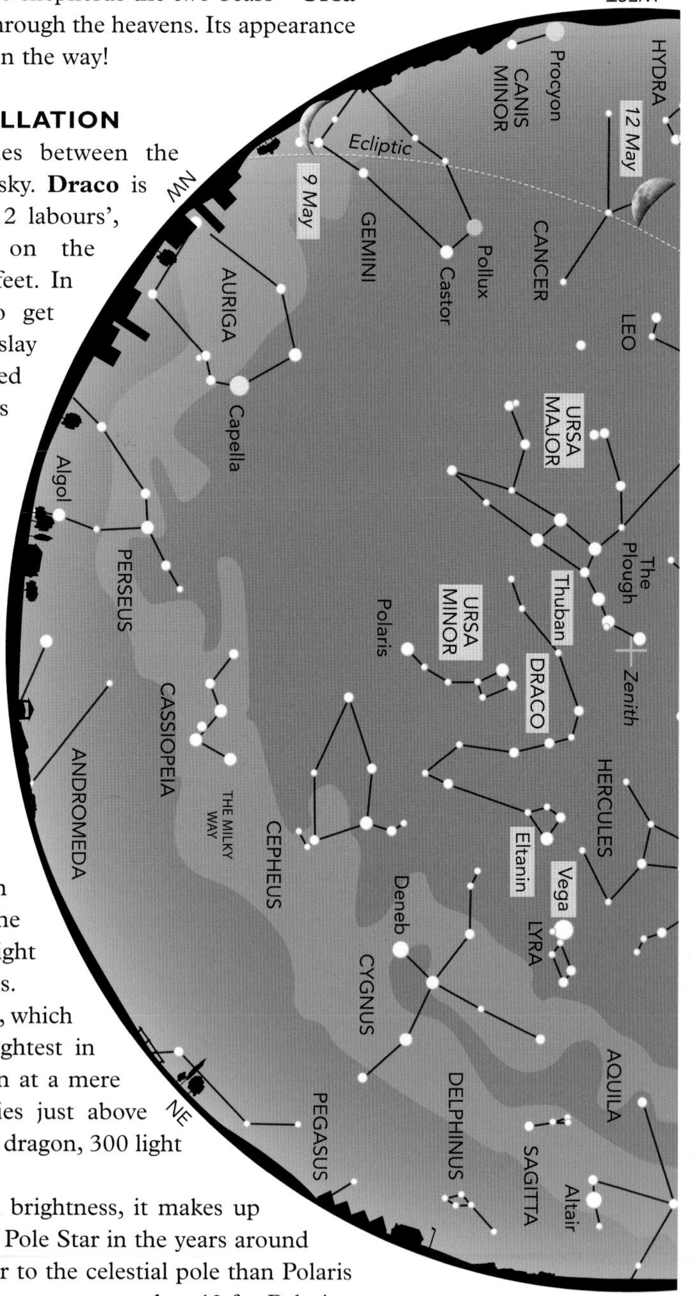

MAY'S CONSTELLATION

The cosmic dragon writhes between the two bears in the northern sky. **Draco** is associated with Hercules' '12 labours', because its head rests on the (upside-down) superhero's feet. In this case, Hercules had to get past a crowd of nymphs and slay a 100-headed dragon (called Ladon) before completing his task – which was the act of stealing the immortal golden apples (guarded by Ladon) from the gardens of the Hesperides.

The brightest star in Draco, confusingly, is gamma Draconis – which ought to be third in the pecking order. Also known as **Eltanin**, this orange star shines at magnitude +2.2, and lies 148 light years away. But all this is to change. In 1.5 million years, it will swing past the Earth at a distance of 28 light years – outshining even Sirius.

Alpha Draconis (**Thuban**), which by rights should be the brightest in the constellation, stumbles in at a mere magnitude +3.7. Thuban lies just above Ursa Minor, in the tail of the dragon, 300 light years away.

But what Thuban lacks in brightness, it makes up for in fame. Thuban was our Pole Star in the years around 2800 BC. It actually lay closer to the celestial pole than Polaris does now – just 2.5 arc minutes, as opposed to 42 for Polaris.

May, and 10 pm at the end of the month. The planets move slightly relative to the stars during the month.

The swinging of Earth's axis – like the toppling of a spinning top – takes place over a period of 26,000 years. 'Precession' means that the Earth's north pole points to a number of stars over the millennia, which we spin underneath – so they appear stationary in the sky. Look forward to AD 14,000, when brilliant **Vega** will take over the pole! It will be Thuban's turn again in AD 20,346.

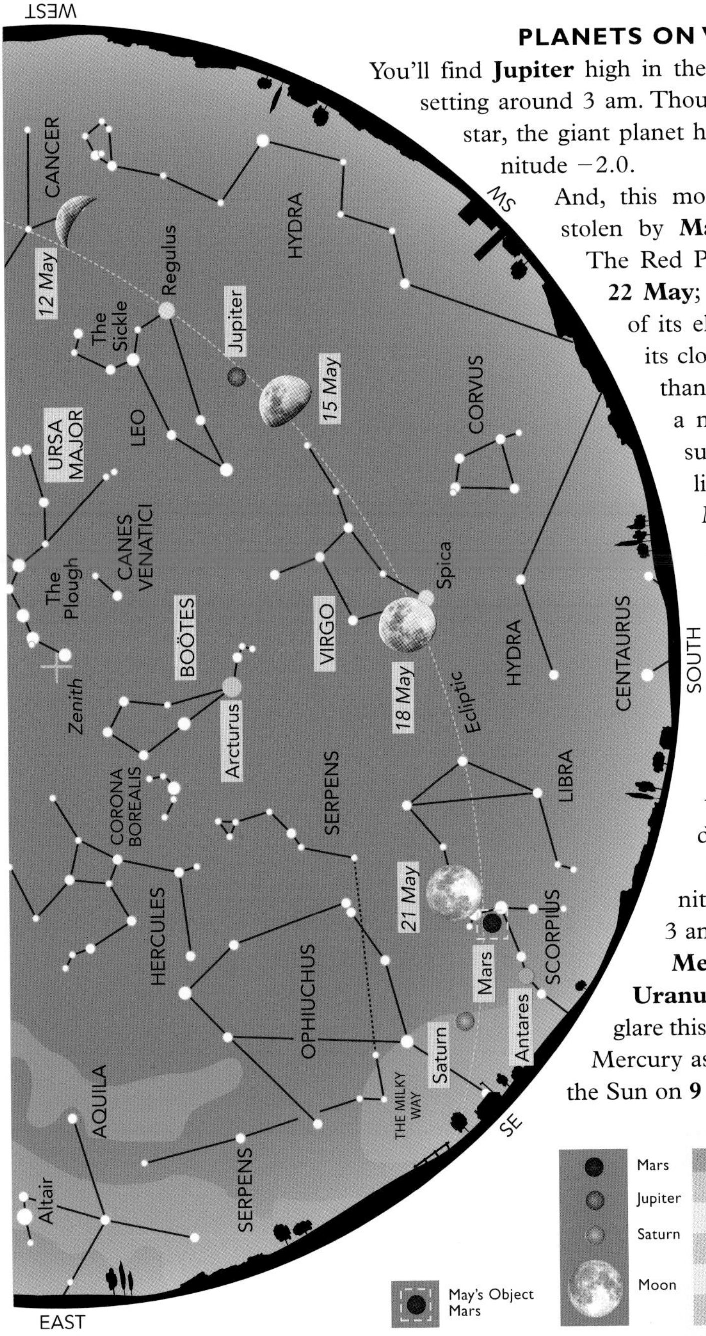

PLANETS ON VIEW

You'll find **Jupiter** high in the south-west, in Leo, and setting around 3 am. Though still brighter than any star, the giant planet has faded slightly, to magnitude −2.0.

And, this month, Jupiter's thunder is stolen by **Mars** (see May's Object). The Red Planet is at opposition on **22 May**; eight days later (because of its elliptical orbit), Mars is at its closest to the Earth – nearer than it's been since 2005. With a magnitude of −2.1, Mars surpasses even Jupiter's brilliance. Lying in Scorpius, Mars rises in the south-east as the sky grows dark, and is visible all night long.

The bright 'star' to the left of Mars is **Saturn**, shining at magnitude +0.2 in Ophiuchus; the ringed planet, too, is above the horizon throughout the hours of darkness.

Dim **Neptune**, magnitude +7.9, rises around 3 am and lies in Aquarius.

Mercury, **Venus** and **Uranus** are hidden in the Sun's glare this month – but watch out for Mercury as it passes bang in front of the Sun on **9 May** (see Special Events)!

MOON		
Date	Time	Phase
6	8.29 pm	New Moon
13	6.02 pm	First Quarter
21	10.14 pm	Full Moon
29	1.12 pm	Last Quarter

MOON

The First Quarter Moon lies near Regulus on **13 May**; on **14 May**, it's between Regulus (right) and brilliant Jupiter (left); and on **15 May**, you'll find the Moon to the left of Jupiter. The star near the Moon on **18 May** is Spica. The Full Moon on **21 May** passes directly over Mars; and it sails above Saturn on **22 May**.

SPECIAL EVENTS

Shooting stars from the Eta Aquarid meteor shower – tiny pieces shed by Halley's Comet burning up in Earth's atmosphere – fly across the sky on the night of **5/6 May**. This year is great for observing these shooting stars, as the Moon is well out of the way.

On **9 May**, we're treated to unusual transit of Mercury (see this month's Topic and Picture).

Mars is at opposition on **22 May** (see May's Object).

MAY'S OBJECT

On **22 May, Mars** is at opposition – in line with the Sun and Earth. Its magnitude reaches −2.1 in the constellation of Scorpius. It's close to the star **Antares** – 'the rival of Mars' – so it's a fascinating time to compare the redness of these two ruddy objects. Use a small telescope to skim over Mars' mottled surface, and spot its icy polar caps.

The debate about ice, water and life on Mars has hotted up over the last few years. There's evidence from NASA's Viking landings in 1976 that primitive bacterial life exists there. And the present flotilla of space probes, which are crawling over its surface or orbiting the Red Planet, are unanimously picking up evidence for present or past water all over Mars – the essential ingredient for life.

NASA's Curiosity mission, which landed on Mars in August 2012, is actively sniffing around Gale Crater. Its rover – the size of a Mini Cooper car – is investigating the geology, composition and life-potential of the Red Planet. Recently, Curiosity has

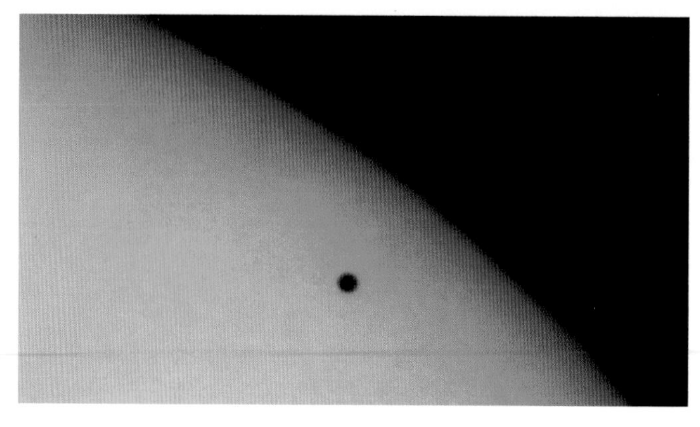

◀ *This image of tiny Mercury transiting across the face of the Sun was taken on 7 May 2003 by Jamie Cooper using a digital camera attached to an ETX90 telescope. Jamie recalls: 'I think I used a primitive digital compact camera fitted against the eyepiece of the scope with an adaptor.' But it shows what great pictures you can get!*

> ***Viewing tip***
> It's best to view your favourite objects when they're well clear of the horizon. If you observe them low down, you're looking through a large thickness of the atmosphere – which is always shifting and turbulent. It's like trying to observe the outside world from the bottom of a swimming pool! This turbulence makes the stars appear to twinkle. Low-down planets also twinkle – although to a lesser extent, because they subtend tiny discs, and aren't so affected.

discovered traces of methane – a gas usually emitted on Earth by microbes – plus soil containing fatty acids, which could be a signature of primitive life.

And this year two more missions – the ExoMars Orbiter and InSight – should be on their way to probe the Red Planet in even more detail. Even normally conservative NASA is admitting its latest missions are designed to pave the way for humans to travel to Mars.

MAY'S PICTURE

The last time we saw a **transit of Mercury** from Britain was in 2003, when much of the country was blessed with fine weather – let's hope for good luck this time!

If you miss this one, though, don't despair. Because Mercury is so close to the Sun, it transits our local star regularly, 13 or 14 times every century (you can expect to see about half of these from any particular location). So, if we're unlucky, you just need to hold out until 11 November 2019....

MAY'S TOPIC
Transit of Mercury

On **9 May**, we're treated to the sight of our innermost planet crossing the face of the Sun. The **transit of Mercury** begins at 0.12 pm; it's halfway across at 3.57 pm, and the planet clears the Sun's disc at 7.42 pm. The planet's silhouette looks like a small, sharp blob – unlike the fuzzy blur of a sunspot.

But before you observe the transit, we need to put in a warning. **DO NOT LOOK AT THE SUN DIRECTLY, EITHER WITH YOUR UNAIDED EYE, OR WITH ANY OPTICAL DEVICE.** The only way to see the transit safely is to project the Sun's image on to a white screen, using a small telescope.

In the past, transits of the innermost planets – Mercury and Venus – were a way of gauging the size of our Solar System. By observing how quickly these worlds crossed the disc of our local star, astronomers could work out how the planets moved – and it led to them calculating the gravitational dynamics of our family in space.

It was Pierre Gassendi, on 7 November 1631, who first observed a transit of Mercury from France. Next was on the occasion of the coronation of King Charles II, on 3 May 1661: it was seen by Christiaan Huygens from London.

Edmond Halley was also on to the case. He got to see a transit of Mercury on one of his many sea voyages – on this occasion, 7 November 1677, in St Helena.

As the Sun reaches its highest position over the northern hemisphere, giving us the longest days and the shortest nights, June isn't the best month for stargazing. But we have some brilliant planets on view. And you can take advantage of the soft, warm weather to acquaint yourself with the lovely summer constellations of **Hercules**, **Scorpius**, **Lyra**, **Cygnus** and **Aquila**.

▼ *The sky at 11 pm in mid-June, with Moon positions at three-day intervals either side of Full Moon. The star positions are also correct for midnight at the beginning of*

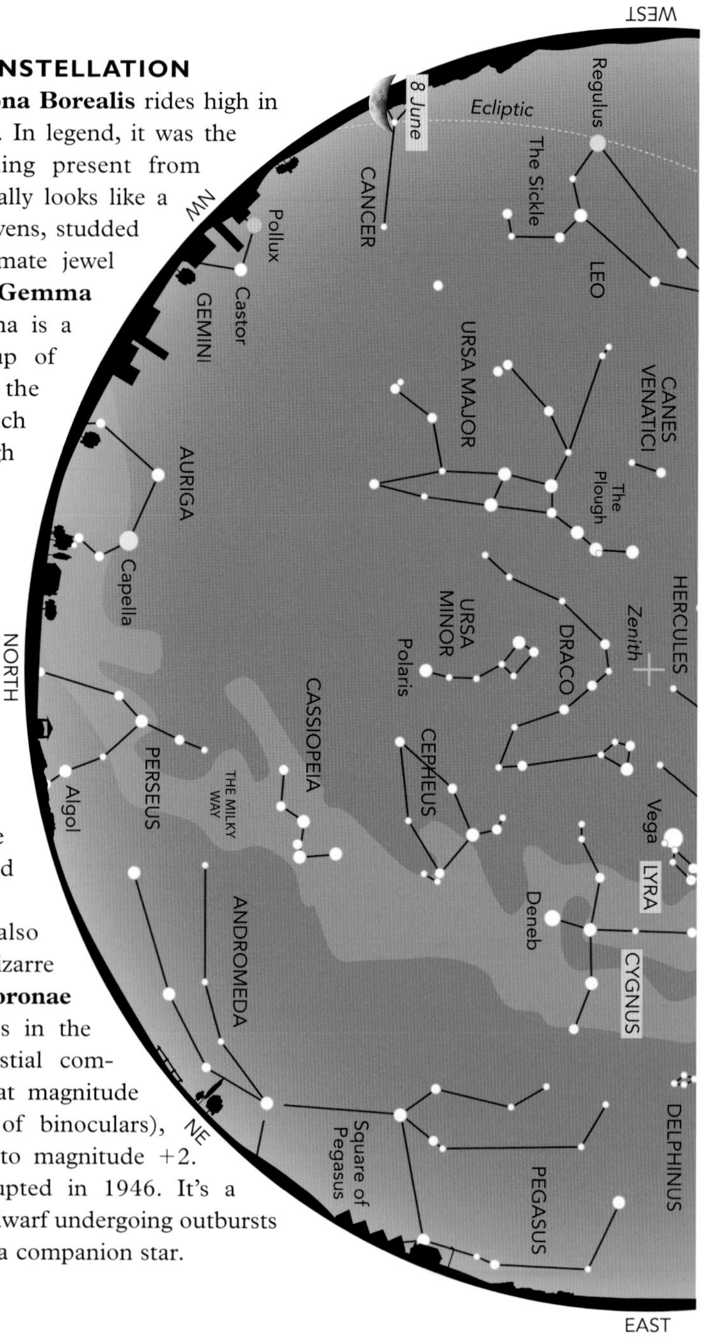

JUNE'S CONSTELLATION

A tiny celestial gem, **Corona Borealis** rides high in the skies of early summer. In legend, it was the crown given as a wedding present from Bacchus to Ariadne. It really looks like a miniature tiara in the heavens, studded at its heart with an ultimate jewel – the blue-white star **Gemma** (magnitude +2.2). Gemma is a member of a large group of young stars (known as the Sirius Supercluster), which all move together through space.

Within the arc of the crown resides **R Coronae Borealis**, a remarkable variable star. It normally hovers around the limits of naked-eye visibility – sixth magnitude – but, unpredictably, it can drop to magnitude +14. That's because dark sooty clouds accumulate above the star's bright surface and obscure its light.

The celestial crown also possesses another rather bizarre variable star, known as **T Coronae Borealis** – which behaves in the opposite way to its celestial compatriot. It usually skulks at magnitude +11 (out of the range of binoculars), and then suddenly flares to magnitude +2. This 'Blaze Star' last erupted in 1946. It's a 'recurrent nova' – a white dwarf undergoing outbursts after dragging material off a companion star.

June, and 10 pm at the end of the month. The planets move slightly relative to the stars during the month.

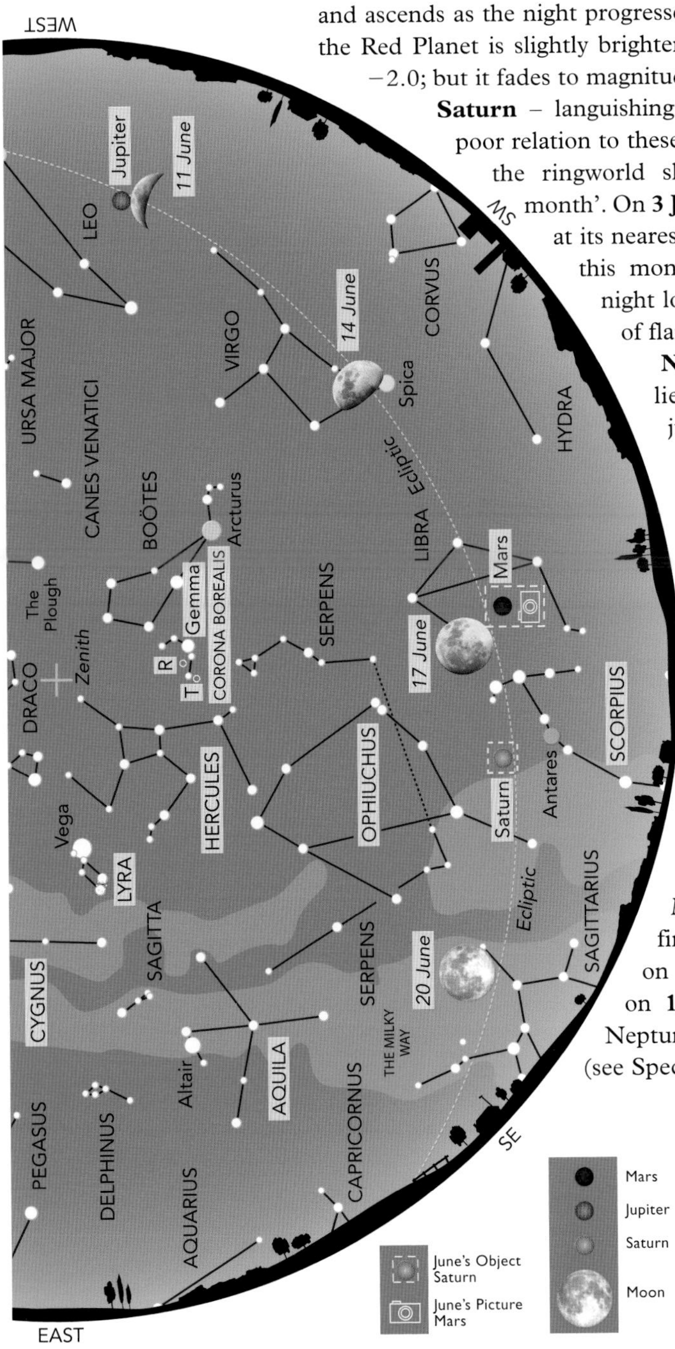

PLANETS ON VIEW

Two brilliant planets are vying for the headlines this month. Shining at a magnificent magnitude −1.8, **Jupiter** lies in Leo – high in the heavens as the sky grows dark – and sets around 1 am.

Mars – in Libra – starts the evening low in the south-east and ascends as the night progresses. At the beginning of June, the Red Planet is slightly brighter than Jupiter, at magnitude −2.0; but it fades to magnitude −1.4 by the month's end.

Saturn – languishing at magnitude +0.1 – is a poor relation to these two show-offs, even though the ringworld should be the 'star of the month'. On **3 June**, Saturn is at opposition: at its nearest and brightest this year (see this month's Object). It's visible all night long in Ophiuchus, to the left of flamboyant Mars.

Neptune (magnitude +7.9) lies in Aquarius, and rises just before 1 am (see Special Events).

Venus and **Uranus** are too close to the Sun to be visible this month; as is **Mercury**, even though it reaches greatest western elongation on **5 June**.

MOON

On **10 June**, the crescent Moon lies near Regulus; it makes a lovely pairing with Jupiter on **11 June**. Spica is the star below the Moon on **14 June**. You'll find the Moon above Mars on **17 June**; it overflies Saturn on **18 June**. The Moon occults Neptune on the night of **25/26 June** (see Special Events).

MOON		
Date	Time	Phase
5	3.59 am	New Moon
12	9.10 am	First Quarter
20	12.02 pm	Full Moon
27	7.18 pm	Last Quarter

◀ *Martin Lewis photographed Mars in 2014, from St Albans. The planet was past opposition and a shadow can be seen on the slopes of the volcano Ascraeus Mons at right. Martin used his home-built 455 mm Dobsonian telescope on an equatorial platform to track Mars for this image made with a ZWO ASI120MC colour webcam. He increased the focal ratio to f/35.5 using a Barlow lens so as to get a large image scale. An atmospheric dispersion corrector removed the red and blue fringing caused by Earth's atmosphere.*

SPECIAL EVENTS

On 3 **June**, **Saturn** is at opposition (see June's Object).

20 June, 11.34 pm: Summer Solstice. The Sun reaches its most northerly point in the sky, so 20 June is Midsummer's Day, with the longest period of daylight. Correspondingly, we have the shortest nights.

With a telescope and a clear east horizon, watch the moonrise at around 0.20 am on **26 June**. Over the next half hour, you'll see the rare sight of Neptune emerging from behind the Moon: the event is best observed from eastern regions of the UK.

JUNE'S OBJECT

The slowly-moving ringworld **Saturn** is currently livening up the sprawling constellation of **Ophiuchus** (the Serpent-Bearer). The planet is famed for its huge engirdling appendages: its rings would stretch nearly all the way from the Earth to the Moon. Saturn is a glorious sight through a small telescope: the world looks surreal, like an exquisite model hanging in space.

And the rings are just the beginnings of Saturn's larger family. It has at least 62 moons, including Titan – which is also visible through a small telescope. The international Cassini–Huygens mission has discovered lakes of liquid methane and ethane on Titan, and possibly active volcanoes.

Citizen Science: Baby stars blowing bubbles

Track down where new stars are being born in the Milky Way, by eyeballing infrared pictures from the Spitzer Space Telescope – converted into glorious false colour – and searching out bubbles blown by energetic young stars. Take part in the Milky Way Project and you'll also catalogue unknown star clusters and unexpected odd objects. In 2015, volunteers found mystery 'yellow balls': hot spheres of gas and dust around the youngest stars, that professional astronomers hadn't spotted before – and hadn't even predicted! **http://www.milkywayproject.org/**

Viewing tip

June is *the* month for the best Sun-viewing. This year our local star has just passed the peak of its 11-year cycle of magnetic activity, and it should still be displaying a fair number of sunspots, with their associated flares and eruptions. But be careful. **NEVER** look at the Sun directly, with your naked eyes or – especially – with a telescope or binoculars: it could well blind you permanently. Fogged film is no safer, because it allows the Sun's infrared (heat) rays to get through. Eclipse glasses are safe (unless they're scratched). The best way to observe the Sun is to project its image through binoculars or a telescope on to a white piece of card. Or – if you want the real 'biz' – get some solar binoculars, with filters that guarantee a safe view. Check the web for details.

And the latest exciting news is that Cassini has imaged plumes of salty water spewing from its icy moon Enceladus, while Dione has traces of oxygen in its thin atmosphere. These discoveries raise the intriguing possibility of primitive life on Saturn's moons....

Saturn itself is second only to Jupiter in size. But it's so low in density that were you to plop it in an ocean, it would float. Like Jupiter, Saturn has a ferocious spin rate – 10 hours and 32 minutes – and its winds roar at speeds of up to 1800 km/h.

Saturn's atmosphere is much blander than that of its larger cousin. But it's wracked with lightning-bolts 1000 times more powerful than those on Earth.

JUNE'S PICTURE

Mars, the Red Planet, is still brilliant in our skies. This image, taken when Mars was close to Earth in 2014, shows a close-up of our neighbour world. The markings on Mars – which were once thought to be vegetation – are now known to be expanses of dark rocks. Like the Earth, Mars has polar caps, as you can see in this image: they're made of frozen carbon dioxide and ice. It's rumoured that a private space company may be sending humans on a one-way trip to the Red Planet in a few years' time: but our recommendation is that you don't hold your breath!

JUNE'S TOPIC
Noctilucent clouds

Look north at twilight, and you may be lucky enough to see what has to be the most ghostly apparition in the night sky – **noctilucent clouds**. Derived from the Latin 'night shining', these spooky clouds glow blue-white. Illuminated by the Sun from below the horizon, they're most commonly seen between latitudes 50° and 70° in the summer, when the Sun towers over the northern hemisphere.

These are the highest clouds in the sky, occurring around 80 kilometres up in the atmosphere. And their origin is controversial. They're certainly composed of ice, coated around tiny particles of dust – but what is the nature of the dust?

Tellingly, the first observation of noctilucent clouds was made in 1885, two years after the eruption of Krakatoa. So could the particles be volcanic dust? Others believe that the dust could be micrometeorites, entering the atmosphere at high altitudes. Some scientists put them down to the industrial revolution, with its resultant increased pollution.

The brilliant trio of the **Summer Triangle** – the stars **Vega, Deneb** and **Altair** – is composed of the brightest stars in the constellations **Lyra, Cygnus** and **Aquila**. And this is the time to catch the far-southern constellations of **Sagittarius** and **Scorpius** – embedded in the glorious heart of the Milky Way.

▼ *The sky at 11 pm in mid-July, with Moon positions at three-day intervals either side of Full Moon. The star positions are also correct for midnight at the beginning of*

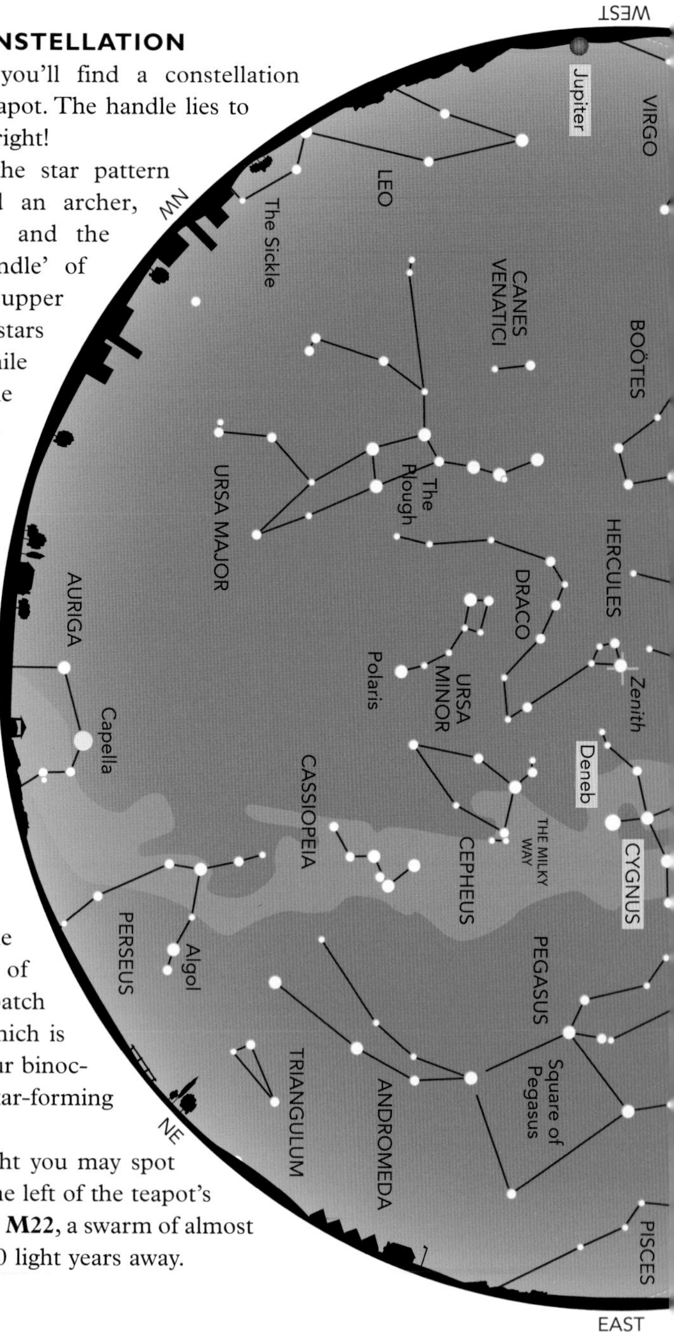

JULY'S CONSTELLATION

Low down in the south, you'll find a constellation that's shaped rather like a teapot. The handle lies to the left and the spout to the right!

To the ancient Greeks, the star pattern of **Sagittarius** represented an archer, with the torso of a man and the body of a horse. The 'handle' of the teapot represents his upper body; the curve of three stars to the right his bent bow; while the end of the spout is the point of the arrow, aimed at Scorpius, the fearsome celestial scorpion.

Sagittarius is rich in nebulae and star clusters. If you have a clear night (and preferably from a southern latitude), sweep Sagittarius with binoculars for some fantastic sights. Above the spout lies the wonderful **Lagoon Nebula** – a region of starbirth that's visible to the naked eye on clear nights (see July's Object). Between the teapot and the neighbouring constellation of Aquila, you'll find a bright patch of stars in the Milky Way, which is catalogued as **M24**. Raise your binoculars higher to spot another star-forming region, the **Omega Nebula**.

Finally, on a very dark night you may spot a fuzzy patch, above and to the left of the teapot's lid. This is the globular cluster **M22**, a swarm of almost a million stars that lies 10,600 light years away.

July, and 10 pm at the end of the month. The planets move slightly relative to the stars during the month.

PLANETS ON VIEW

Jupiter shines brilliantly in the west after sunset, at magnitude −1.7. Lying in Leo, the giant planet is setting about 11 pm.

As the Earth draws away from **Mars**, the Red Planet is fading dramatically: from magnitude −1.4 to −0.8 during July. You'll find Mars in Libra, setting around 1 am.

Saturn (magnitude +0.3) follows Mars below the horizon about 2 am: it lies in Ophiuchus, above Scorpius' brightest star, Antares.

Binocular-world **Neptune** is lurking in Aquarius; at magnitude +7.8, it rises around 11 pm.

Uranus (magnitude +5.8) begins to become visible this month, rising about midnight in Pisces.

Mercury and **Venus** are hidden in the Sun's glare this month.

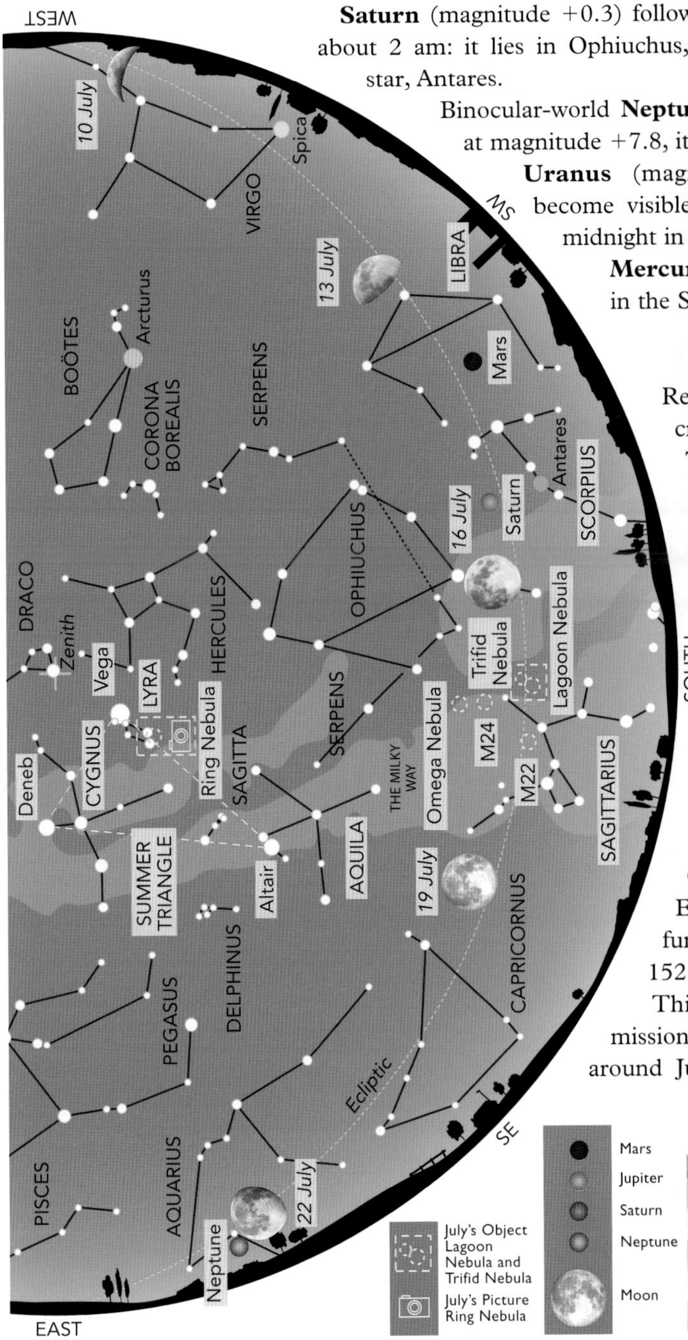

MOON

Regulus lies above the crescent Moon on 7 **July**. The brilliant object near the Moon on **8 and 9 July** is Jupiter. The Moon passes over Mars on **14 July**. On **15 July**, you'll find Saturn to the left of the Moon. On the morning of **29 July**, the crescent Moon lies near Aldebaran and the Hyades.

SPECIAL EVENTS

On **4 July**, at 5.24 pm, the Earth reaches aphelion, its furthest point from the Sun – 152 million kilometres out.

This month, NASA's Juno mission is due to go into orbit around Jupiter. It will examine the

MOON		
Date	Time	Phase
4	12.01 pm	New Moon
12	1.52 am	First Quarter
19	11.56 pm	Full Moon
26	11.59 pm	Last Quarter

◀ *Ed Cloutman captured this image of the Ring Nebula on 19 September 2012. Observing from South Wales, he used an IKHARUS 10-inch (250 mm) Ritchey-Chretien telescope. Ed took the image using an Astro Physics 900GTOCamera, with Astronomik RGB and H-alpha filters: the total exposure was 1 hour 20 minutes.*

planet's stormy weather, the composition of its 'air' and also its intense gravity and magnetic fields, for clues to the deep interior of the giant of the Solar System.

JULY'S OBJECT

The magnificent constellation of Sagittarius – alas, too far south to be a sensational sight from Britain – is home to a pair of glorious nebulae. These gaseous crucibles of starbirth are among the most heavenly sights in the sky. The **Lagoon Nebula** is an oasis of calm: this gentle womb of burgeoning stars is home to many 'Bok globules' – small black clouds hatching baby suns.

Its companion, the **Trifid Nebula**, looks dramatically different. It's dissected by bands of dark cosmic dust – interstellar soot – which is poised to collapse and create fledgling stars. NASA's Spitzer orbiting telescope, which looks at the Universe in infrared (heat rays), has discovered 120 newborn stars in the nebula. The Trifid lies just over 5000 light years away.

Citizen Science: Make contact with alien life

Tune into the largest radio telescope in the world – the 305-metre dish nestled in the green jungle of Puerto Rico – to search for signals from extraterrestrial civilizations. After all, we broadcast to the inhabitants of our planet, so why won't ET want to talk to us? The Arecibo telescope is used for conventional radio astronomy: but piggybacked on to its receiver is a device that can listen into artificial signals from space. You need to log into SETI@home on your computer to help the astronomers at the University of California, Berkeley – who are putting out the signals on the web – and you could be the person who first hears from an alien! **http://setiathome.ssl.berkeley.edu/**

JULY'S PICTURE

The **Ring Nebula** in **Lyra** looks like a celestial smoke ring. It appears the same size as Jupiter in the sky, and William Herschel – who discovered Uranus – found many similar puzzling objects. They looked to him to be very similar to the planet he had stumbled over. So he called them 'planetary nebulae'.

But the Ring Nebula is a dying star – a fate that will happen to our Sun in some 7 billion years' time. Its core has run out of nuclear fuel, and the unstable star has puffed off its outer layers into space. Eventually, these layers will disperse, leaving the core (centre) exposed as a cooling white dwarf star – which will later become a black, celestial cinder.

Lying between the two lowest stars of the tiny constellation, the Ring Nebula is very faint – around magnitude +9. It's best to have a telescope with a mirror of around 200mm to observe the planetary nebula well.

Viewing tip

This is the month when you really need a good, unobstructed horizon to the south, for the best views of the glorious summer constellations of Scorpius and Sagittarius. They never rise high in temperate latitudes, so make the best of a southerly view – especially over the sea – if you're away on holiday. A good southern horizon is also best for views of the planets, because they rise highest when they're in the south.

JULY'S TOPIC
Life in the Solar System

Controversial though it may be, we believe that the evidence for life on Mars is indisputable. In 1976, the twin Viking probes landed on the Red Planet. There were four experiments on board, designed specifically to look for traces of life. One experiment proved positive. No: it wasn't proof of 'little green men', but more of the existence of 'little green slime'.

And this is how astrobiologists generally view the prospects of life in the Universe. It may have been created: but did it evolve? On Earth, we had an astonishing explosion in evolution: from single-celled amoebae to the dinosaurs; then mammals and humankind. What next?

Are there other worlds in the Solar System that could harbour life? The top of the tree is Saturn's moon Titan, which boasts oceans of ethane and methane – packed with organic chemicals – under a nitrogen atmosphere, like that of the Earth.

Then there's Jupiter's moon Europa – a frozen world whose ices may cover a warm, submerged ocean in which alien fish might swim. And Saturn's moon Enceladus has warm and active volcanoes: just the kind of place where life feels at home.

The planets are undergoing some breathtaking choreography this month. Plus: we hope there's going to be an unusually spectacular display of shooting stars, after midnight on **11/12 August** – worth organizing a late-night **Perseid** star party!

▼ *The sky at 11 pm in mid-August, with Moon positions at three-day intervals either side of Full Moon. The star positions are also correct for midnight*

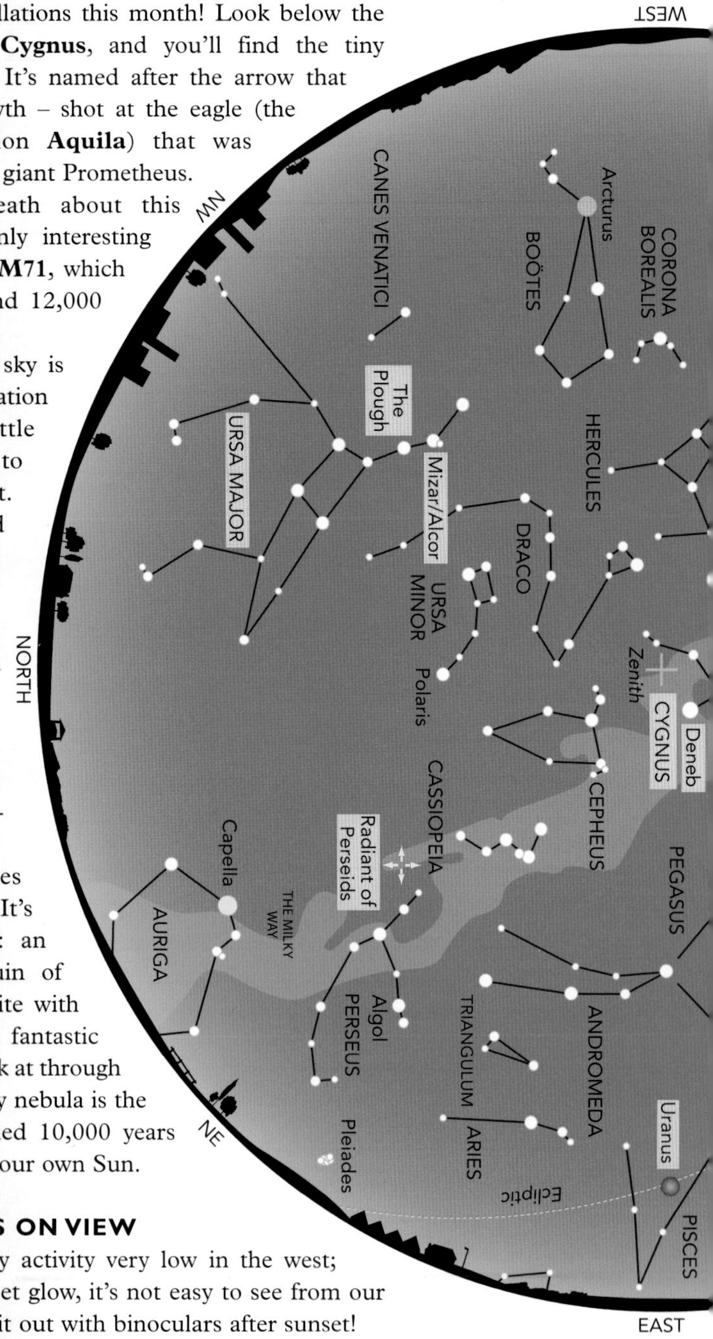

AUGUST'S CONSTELLATION

Two very obscure constellations this month! Look below the great celestial cross of **Cygnus**, and you'll find the tiny constellation of **Sagitta**. It's named after the arrow that **Hercules** – in Greek myth – shot at the eagle (the neighbouring constellation **Aquila**) that was pecking at the liver of the giant Prometheus.

Don't hold your breath about this gathering of stars: its only interesting object is the star cluster **M71**, which is 10 billion years old and 12,000 light years away.

Next to Sagitta in the sky is the even fainter constellation of **Vulpecula** (the Little Fox), its stars too faint to be marked on our chart. Johannes Hevelius created this constellation in the 17th century, as 'the little fox and goose'. The bird has now flown – but the constellation has acquired fame as the place in the sky where the first pulsar (a rotating neutron star) was discovered in 1967.

The constellation does boast one celestial glory. It's the **Dumbbell Nebula**: an absolutely sensational ruin of a star. It's a huge favourite with amateur astronomers – a fantastic object to image and to look at through a telescope. This planetary nebula is the remains of a star that died 10,000 years ago: a fate that will await our own Sun.

PLANETS ON VIEW

There's a lot of planetary activity very low in the west; unfortunately, in the sunset glow, it's not easy to see from our northern latitude. Check it out with binoculars after sunset!

at the beginning of August, and 10 pm at the end of the month. The planets move slightly relative to the stars during the month.

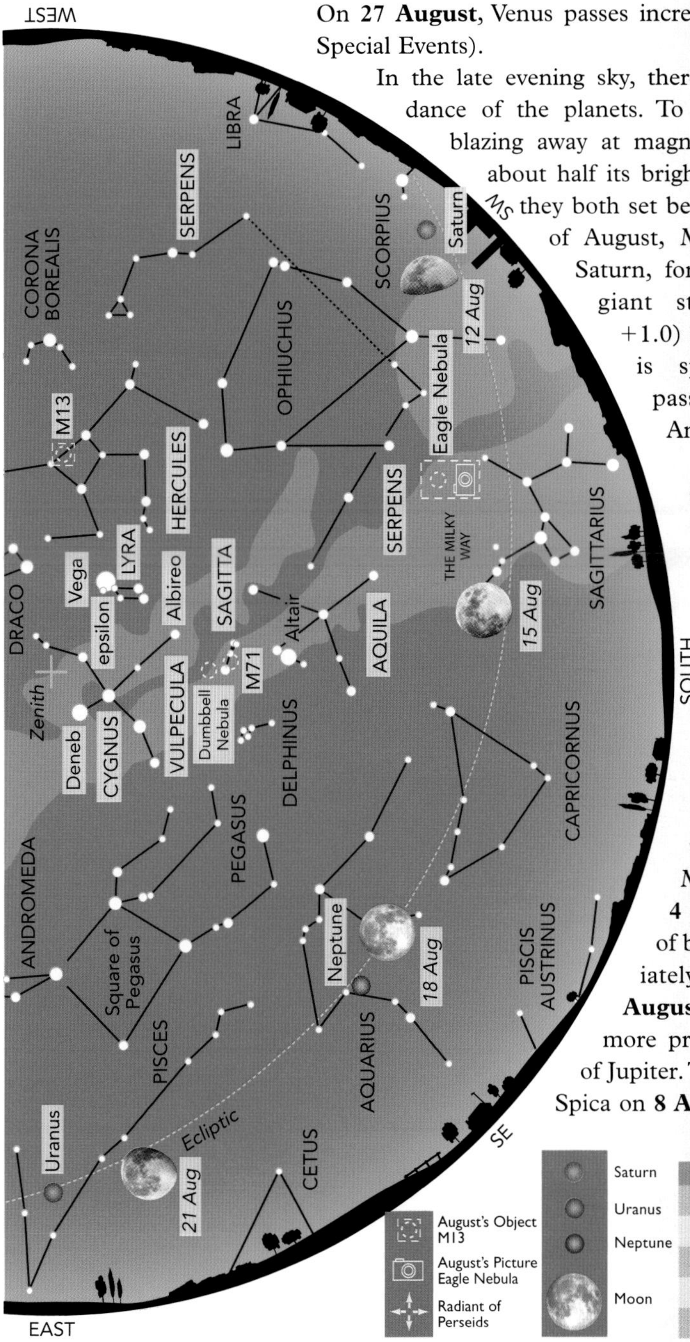

The month opens with **Venus** (magnitude −3.8) hugging the horizon; some 30° to its left, **Jupiter** rides rather higher in Leo, at magnitude −1.6. Between them lies fainter **Mercury**, difficult to spot at magnitude 0.0.

As the month progresses, the three planets converge: Mercury reaches greatest eastern elongation on **16 August**. On **27 August**, Venus passes incredibly close to Jupiter (see Special Events).

In the late evening sky, there's a much more obvious dance of the planets. To the south-west, **Mars** is blazing away at magnitude −0.5, with **Saturn** about half its brightness at magnitude +0.5: they both set before midnight. At the start of August, Mars lies to the right of Saturn, forming a triangle with red giant star Antares (magnitude +1.0) below. But the Red Planet is speeding leftwards, and passes between Saturn and Antares on **24 August**.

The dim planet **Neptune** (magnitude +7.8) lies in Aquarius and rises at 9 pm; slightly brighter **Uranus** emerges above the horizon at 10 pm, at magnitude +5.8 in Pisces.

MOON

Here's a lunar challenge: with binoculars, spot the very narrowest crescent Moon in the dusk on **4 August**, lying to the left of brilliant Venus, and immediately below Mercury. On **5 August**, the crescent Moon is more prominent, just to the right of Jupiter. The waxing Moon is above Spica on **8 August**. On **11 August**, the

MOON		
Date	Time	Phase
2	9.44 pm	New Moon
10	7.21 pm	First Quarter
18	10.26 am	Full Moon
25	4.41 am	Last Quarter

Moon forms a striking triangle with Saturn (left) and Mars (lower left). It's to the upper left of Saturn on **12 August**. The bright star near the waning Moon on the mornings of **25 and 26 August** is Aldebaran.

SPECIAL EVENTS

The maximum of the **Perseid meteor shower** falls on **11/12 August**, when the Earth runs into debris from Comet Swift–Tuttle. Always a reliable and prolific shower, the Perseids may be even more spectacular this year as we encounter streams of dust ejected from the comet in 1479 and 1862. The best views will be after midnight, when the Moon has set.

On **27 August**, the two brightest planets are only 9 arc minutes apart, as Venus passes Jupiter. From Britain, the conjunction is very low in the evening twilight; but it's a stunning sight as seen from Brazil.

AUGUST'S OBJECT

At the darkest part of an August night, you may spot a faint fuzzy patch way up high in the south-west. Through binoculars, it appears as a gently glowing ball of light. With a telescope, you can glimpse its true nature: a cluster of a third of a million stars, swarming together in space.

This wonderful object is known as **M13**, because it was the 13th entry in the catalogue of fuzzy objects recorded by the 18th-century French astronomer Charles Messier. We now classify M13 as a 'globular cluster'. These great round balls of stars are among the oldest objects in our Milky Way Galaxy, dating back to its birth some 13 billion years ago.

◀ Within the glowing Eagle Nebula (M16), you can spot the dark 'pillars of creation' in this image from Ed Cloutman taken on 9 July 2013 from South Wales. The total exposure time was 2 hours, with an Astro Physics 900GTOCamera mounted on an IKHARUS 10-inch (250 mm) Ritchey-Chretien telescope. Ed used Astronomik LRGB and H-alpha filters; digital processing included 'fat-tail deconvolution' to sharpen the image.

Viewing tip

Have a meteor party to check out the Perseid meteor shower on **11/12 August**! You don't need any optical equipment – in fact, telescopes and binoculars will restrict your view of the meteor shower. The ideal viewing equipment is your unaided eye, plus a sleeping bag and a lounger on the lawn. If you want to record your observations, take a watch, notepad and torch, and note the times when meteors appear. Try to observe for at least an hour.

In 1974, radio astronomers sent a message towards M13, hoping to inform the inhabitants of any planet there of our existence. There's only one problem: M13 lies so far away that we wouldn't receive a reply until AD 14,300!

AUGUST'S PICTURE

The constellation of **Serpens** (the Serpent) has very little to write home about, except for consisting of two disconnected halves. But it does boast a gem, in the shape of the **Eagle Nebula** (M16) – famed for a sensational image from the Hubble Space Telescope. This picture revealed the glory of its star-forming regions: the fantastic 'pillars of creation' that have gone down as a legend in cosmic sky-sights.

The Eagle Nebula was discovered by Jean-Philippe de Cheseaux in 1745. The burgeoning nebula is 7000 light years away, and is home to nearly 500 infant stars. While our Sun boasts an age of nearly 5 billion years, the toddlers of the Eagle Nebula are a mere 5.5 million years old.

AUGUST'S TOPIC
Double stars

The Sun is an exception in having singleton status. Over half the stars you see in the sky are paired up – they're double stars, or binaries. Look no further than the **Plough** in **Ursa Major** for a beautiful naked-eye example: the penultimate star in the Plough's handle (at the bend) is clearly double. **Mizar** (the brighter star) and its fainter companion **Alcor** have often been named 'the horse and rider'.

Another naked-eye double lies in **Lyra**. Next to the brilliant white star **Vega** is **epsilon Lyrae**, which is visible to the keen-sighted as a pair of stars. Use a small telescope, and you'll discover that the family has grown: epsilon Lyrae is actually a double-double!

The four stars are about 162 light years away, and may form part of a bigger system: there are hints that up to ten stars could be involved.

Move next to the neighbouring constellation of **Cygnus**. The huge, cross-shaped constellation is dominated by **Deneb**, which represents the celestial flying swan's tail. But look to the end of the swan – its head – and you'll discover a fainter star named **Albireo**. It's the most beautiful double star in the Universe: but you'll definitely need a telescope for this one. Lying about 430 light years away, the two stars make an amazing colour contrast – the cooler, yellow star, paired with a hot blue star. The colours are really vivid: we recommend Albireo as a cosmic must-see.

This month, the nights become longer than the days, as the Sun migrates southwards in the sky. Autumn is here – with its unsettled weather – and we have wet star patterns to match! **Aquarius** (the Water Carrier) is part of a group of aqueous star patterns which include **Cetus** (the Sea Monster), **Capricornus** (the Sea Goat), **Pisces** (the Fishes), **Piscis Austrinus** (the Southern Fish) and **Delphinus** (the Dolphin).

▼ *The sky at 11 pm in mid-September, with Moon positions at three-day intervals either side of Full Moon. The star positions are also correct for midnight at*

SEPTEMBER'S CONSTELLATION

The flying swan – **Cygnus** – is one of our most cherished constellations. The celestial bird actually looks like its namesake, with outspread wings and an elongated neck.

Swan legends abound. One of the most popular is that Zeus – disguised as a swan – seduced Leda, the wife of King Tyndareus of Sparta. As a result, the unfortunate woman gave birth to twins, one immortal and one mortal: they appear in the sky as Pollux and Castor, the heavenly twins in the constellation of Gemini.

Deneb forms the swan's tail. It's the furthest away of the top 20 brightest stars, but its distance is hard to pin down. Estimates range from 1500 to 2600 light years, which means that the star shines anything between 50,000 and 200,000 times brighter than the Sun.

The swan's head is marked by what's probably the most beautiful double star in the sky: **Albireo**. Go to August's Topic for more details.

Another Cygnus gem is the **North America Nebula**. Looking uncannily like its terrestrial lookalike, this red cloud of gas is bigger than the Full Moon; but it's so faint that you'll need binoculars to pick it out.

The Milky Way meanders through Cygnus. Edge-on, the band of our Galaxy is riddled with star clusters and nebulae. It's glorious to sweep Cygnus with binoculars or a small telescope.

the beginning of September, and 10 pm at the end of the month. The planets move slightly relative to the stars during the month.

PLANETS ON VIEW

In the west, **Venus** is gradually moving upwards in the dusk twilight; shining at magnitude −3.8, the brilliant Evening Star sets at 8 pm.

You'll find **Mars** and **Saturn** low in the evening sky, to the south-west: both lie in Ophiuchus, and set around 10 pm. At magnitude −0.1, Mars lies to the left of the rather fainter Saturn (magnitude +0.6). The star forming a triangle to the lower right is red giant Antares, the chief luminary of Scorpius.

Neptune (magnitude +7.8), in Aquarius, reaches opposition on **2 September** and is above the horizon all night long (see this month's Object). To add to the excitement, on **15 September** the most distant planet hides behind the Moon (see Special Events).

At magnitude +5.7, **Uranus** lies in Pisces and rises around 8 pm.

During the last week of September, scour the eastern horizon just before dawn to spot **Mercury** putting on its best morning performance of the year, as it brightens from magnitude +0.5 to −0.5. The innermost planet is at greatest western elongation on **28 September**.

Jupiter is too close to the Sun to be easily visible this month.

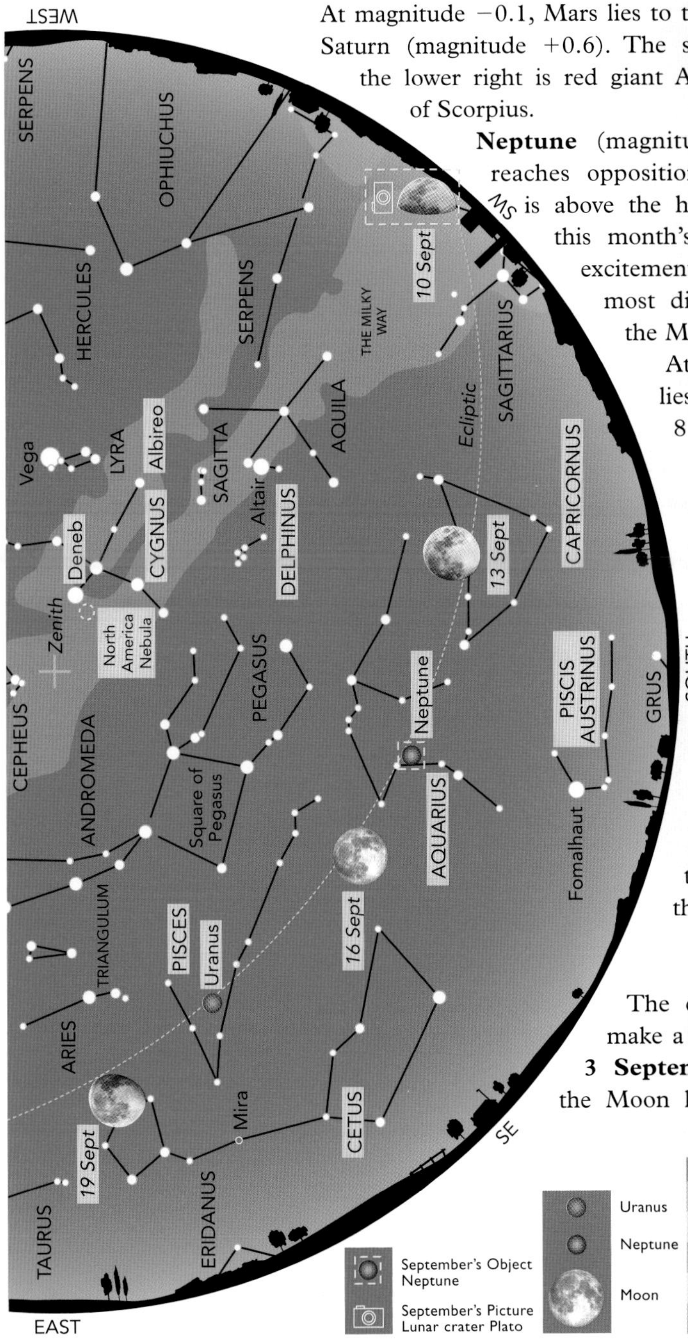

MOON

The crescent Moon and Venus make a lovely sight after sunset on **3 September**. On **8 September**, the Moon lies near Saturn; the First

MOON		
Date	Time	Phase
1	10.03 am	New Moon
9	12.49 pm	First Quarter
16	8.05 pm	Full Moon
23	10.56 am	Last Quarter

Quarter Moon on **9 September** passes well above Mars, with Saturn to the right. The Moon occults Neptune on **15 September** (see Special Events). As the Moon rises on **21 September**, it lies right below Aldebaran. And on the morning of **29 September**, you'll find the narrow crescent Moon just to the right of Mercury.

▲ *The lunar crater Plato, photographed using a 355 mm telescope from Flackwell Heath in Buckinghamshire, by Dave Tyler. Dave used a Celestron C14 SCT working at f/30 with a Flea3 webcam. This is a mosaic of four separate frames in order to cover a wider area of the Moon surrounding the crater.*

SPECIAL EVENTS

People along a narrow band through Africa – from Gabon to Madagascar – are treated to an annular eclipse on **1 September**. The Moon moves directly in front of the Sun but is a tad smaller in size, so you'll see a thin ring (annulus) of the Sun's disc around the Moon's silhouette. The rest of Africa and the Indian Ocean experience a partial solar eclipse. Find more details at eclipse.gsfc.nasa.gov/solar.html.

Neptune is at opposition on **2 September** (see this month's Object). And, on **15 September**, the almost-Full Moon occults Neptune – if you have a telescope, it's an ideal time to locate this dim world. Watch for the most distant planet to emerge from behind the Moon between 8.50 pm and 9.05 pm, depending on your location.

The Autumn Equinox occurs at 3.21 pm on **22 September**. The Sun is over the Equator as it heads southwards in the sky, and day and night are equal.

This month, the InSight lander should reach Mars. It will probe the Red Planet's interior with a seismometer to monitor Mars-quakes, and deploy a 'mole' to dig 5 metres down and measure heat emerging from the planet's depths.

Also in September, NASA is scheduled to launch the OSIRIS-REx mission, to bring home a sample of asteroid Bennu.

SEPTEMBER'S OBJECT

With Pluto being demoted to a mere 'ice dwarf', **Neptune** is officially the most remote planet in our Solar System. It lies 4500 million kilometres out – 30 times the Earth's distance – in the twilight zone of our family of worlds, and takes nearly 165 years to circle the Sun.

Neptune is at its closest this year on **2 September**, and just visible through a small telescope in Aquarius. But you need a space probe to get up close and personal to the gas giant planet. In 1989, Voyager 2 revealed a turquoise world 17 times heavier than Earth, cloaked in clouds of methane and ammonia.

The most distant planet has a family of 14 moons, including Triton, which boasts erupting ice volcanoes. And the world is encircled by very faint rings of dusty debris.

For a world so far from the Sun, Neptune is amazingly frisky. Its core blazes at nearly 5000°C – as hot as the Sun's surface. This internal heat drives dramatic storms, and winds of 2000 km/h – the fastest in the Solar System.

SEPTEMBER'S PICTURE

The **lunar crater Plato** is one of the rare impacts on the Moon that boasts a dark surface. Was it caused by a deep collision, that caused subsurface magma to well up – as in the case of the lunar 'seas'? Being dark, Plato is very prominent on our lunar companion. The 17th-century astronomer Hevelius called it 'the Greater Black Lake'.

The crater is 109 kilometres across and is estimated to be around 4 billion years old. You can locate it near Mare Imbrium, and it's best seen a day or two after First Quarter.

Viewing tip

Don't think that you need a telescope to bring the heavens closer. Binoculars are excellent – and you can fling them into the back of the car at the last minute. For astronomy, buy binoculars with large lenses coupled with a modest magnification. Binoculars are described as being, for instance, '7×50' – meaning that the magnification is seven times, and that the diameter of the lenses is 50 millimetres. These are ideal for astronomy – they have good light grasp, and the low magnification means that they don't exaggerate the wobbles of your arms too much. It's always best to rest your binoculars on a wall or a fence to steady the image. Some amateurs are the lucky owners of huge binoculars – say, 20×70 – with which you can see the rings of Saturn (being so large, these binoculars need a special mounting). But above all, *never* buy binoculars with small lenses that promise huge magnifications – they're a total waste of money.

SEPTEMBER'S TOPIC
Zodiacal light

At this time of year (and also in spring) – the Zodiac hangs high in the sky. Forget the rubbish of astrology: in astronomy, this is the path that the planets pursue in our Solar System. And now is the time to observe one of its most elusive phenomena: the zodiacal light.

On a really clear night, away from streetlights, you may spot a faint pyramid of light in the west just after the Sun has gone down. This ghostly glow is the zodiacal light.

It's also visible in the morning sky. The 12th-century poet and astronomer Omar Khayyam had a fantastic view of this 'false dawn' over the Persian desert. In typical fashion, he celebrated with wine – and wrote a poem: 'When false dawn streaks the east with cold, grey line, pour in your cups the pure blood of the vine.'

It's so rarely seen that many astronomers have never witnessed it. But persevere! It's caused by light reflected from a fog of tiny particles from old comets and asteroids that have broken up. As our planet orbits the Sun, it scoops up around 40,000 tonnes of this space dust every year.

OCTOBER

The glories of October's skies can best be described as 'subtle'. The barren square of **Pegasus** dominates the southern sky, with **Andromeda** attached to his side. But the dull autumn constellations are already being faced down by the brilliant lights of winter, spearheaded by the beautiful star cluster of the **Pleiades**. From Greece to Australia, ancient myths independently describe these stars as a group of young girls being chased by an aggressive male – often **Aldebaran** or **Orion**.

▼ The sky at 11 pm in mid-October, with Moon positions at three-day intervals either side of Full Moon. The star positions are also correct for midnight at

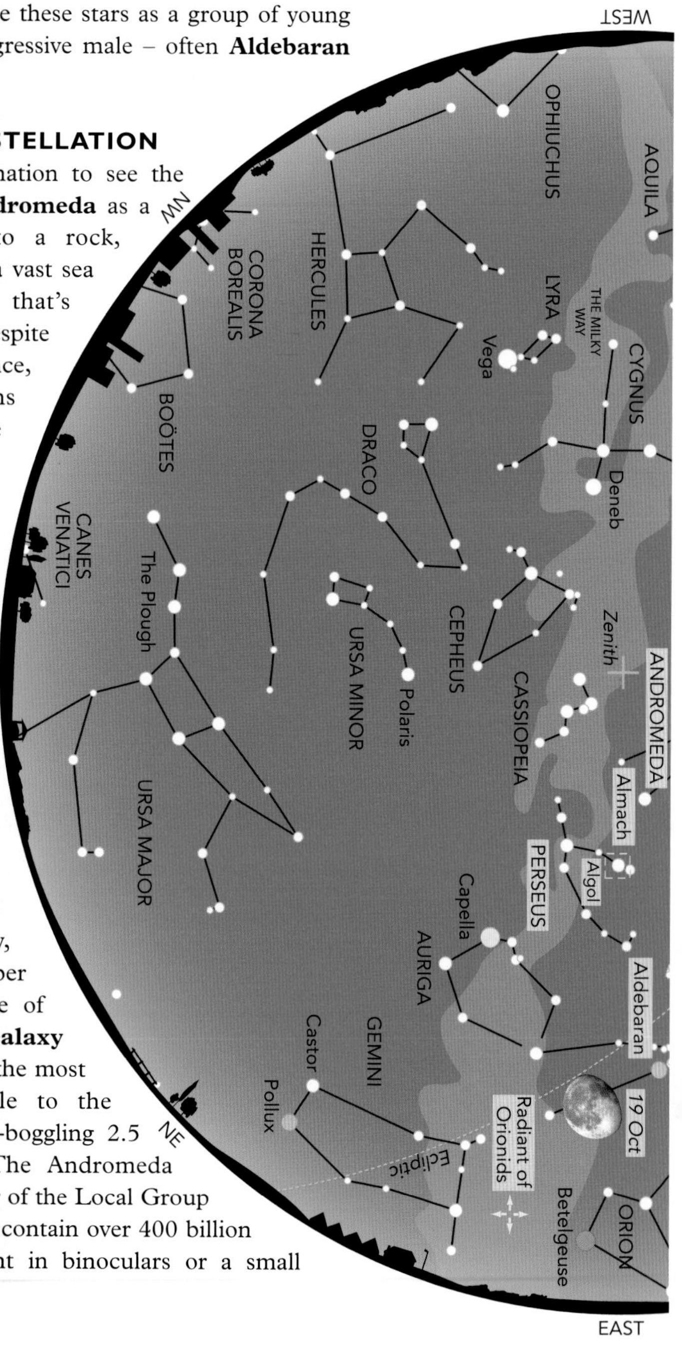

OCTOBER'S CONSTELLATION

It takes considerable imagination to see the line of stars making up **Andromeda** as a young princess chained to a rock, about to be gobbled up by a vast sea monster (**Cetus**) – but that's ancient legends for you. Despite its rather mundane appearance, the constellation contains some surprising delights. One is **Almach**, the star at the left-hand end of the line. It's a beautiful double star. The main star is a yellow supergiant shining 2000 times brighter than the Sun, and its companion – which is fifth magnitude – is bluish. The two stars are a lovely sight in small telescopes. Almach is actually a quadruple star: its companion is in fact triple.

But the real glory of Andromeda is its great galaxy, beautifully placed on October nights. Lying above the line of stars, the **Andromeda Galaxy** (see this month's Picture) is the most distant object easily visible to the unaided eye. It lies a mind-boggling 2.5 million light years away. The Andromeda Galaxy is the biggest member of the Local Group of galaxies – it's estimated to contain over 400 billion stars. It is a wonderful sight in binoculars or a small telescope.

the beginning of October, and 9 pm at the end of the month (after the end of BST). The planets move slightly relative to the stars during the month.

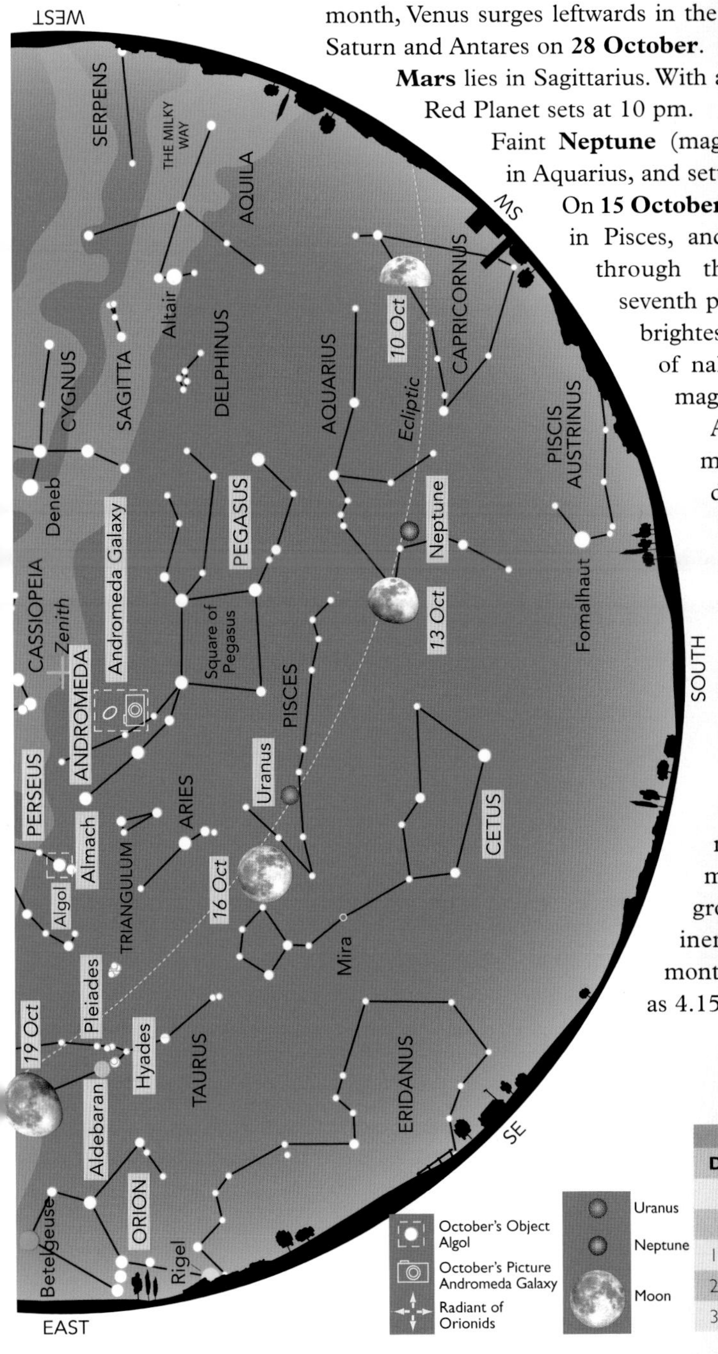

PLANETS ON VIEW

You can't miss **Venus** shining brilliantly in the west after sunset. At magnitude −3.9, the Evening Star sets about 7 pm.

At the beginning of October, **Saturn** lies well to the left of Venus, setting around 8 pm. The ringworld (magnitude +0.6) inhabits Ophiuchus, lying above red giant Antares. During the month, Venus surges leftwards in the sky, and passes between Saturn and Antares on **28 October**.

Mars lies in Sagittarius. With a magnitude of +0.2, the Red Planet sets at 10 pm.

Faint **Neptune** (magnitude +7.8) is lurking in Aquarius, and setting about 3 am.

On **15 October, Uranus** is at opposition in Pisces, and above the horizon all through the night. Though the seventh planet is at its closest and brightest, it's on the borderline of naked-eye visibility, with a magnitude of +5.7.

At the beginning of the month, look low in the dawn sky for a glimpse of **Mercury**, rising at 5.20 am and shining at magnitude −0.8. Gradually sinking lower day by day, Mercury passes close by Jupiter before sunrise on **11 October**.

Giant planet **Jupiter** (magnitude −1.5) is moving upwards in the morning sky in Virgo, and growing ever more prominent. By the end of the month, Jupiter is rising as early as 4.15 am.

MOON		
Date	Time	Phase
1	1.11 am	New Moon
9	5.33 am	First Quarter
16	5.23 am	Full Moon
22	8.14 pm	Last Quarter
30	5.38 pm	New Moon

MOON

On **3 October**, brilliant Venus lies to the lower left of the crescent Moon. You'll find the Moon to the right of Saturn on **5 October**; and to the planet's upper left on **6 October**. The bright reddish 'star' below the Moon on **7 and 8 October** is Mars. On the night of **18/19 October**, the Moon occults the Hyades (see Special Events). Regulus lies just above the waning crescent Moon on the morning of **25 October**. There's a splendid sight just before dawn on **28 October**, when the thin crescent Moon stands right over giant planet Jupiter.

SPECIAL EVENTS

On **15 October**, Uranus is at opposition.

Just after midnight on **18/19 October**, the Moon moves right in front of the **Hyades** star cluster. Depending on your location, you may see up to half a dozen stars of naked-eye brightness being occulted: the Moon's brilliance means you'll need binoculars or a small telescope to view the event, and the stars' reappearance at the dark limb of the Moon will be more spectacular than their disappearance.

Debris from Halley's Comet smashes into Earth's atmosphere on **20/21 October**, causing the annual **Orionid** meteor shower. Unfortunately, this year the shooting stars are largely washed out by bright moonlight.

At 2 am on **30 October**, we see the end of British Summer Time for this year. Clocks go backwards by an hour.

And this month should see the European–Russian ExoMars Trace Gas Orbiter arrive at Mars. From orbit, this mission will scan the Red Planet's atmosphere for gases – like methane – that are produced by living organisms. It will also drop off the probe Schiaparelli, which will test landing systems for the forthcoming ExoMars Rover, designed to ferret out any microbes dwelling in the planet's red deserts.

Citizen Science: Worlds beyond

There are eight planets in our Solar System. But over the past 20 years, astronomers have discovered nearly 2000 planets orbiting other stars. They first picked them up by watching how the parent star 'wobbled' under the gravitational force of its planetary retinue. Space telescopes like Kepler have discovered hundreds of other worlds, by watching how a star's light briefly dims when a planet crosses its disc. These 'extrasolar planets' are a weird bunch: they range from 'hot Jupiters' – gas giants close to their star – to planets as small as the Earth. Join the Planet Hunters project and discover unknown worlds for yourself.
www.planethunters.org

◄ The Andromeda Galaxy, M31, photographed by Ian King using a 75 mm Pentax refractor and Starlight Xpress SXV-M25 CCD camera. He used separate exposures through individual colour filters.

> **Viewing tip**
> The Andromeda Galaxy is often described as the furthest object 'easily visible to the unaided eye'. But it can be a bit elusive – especially if you are suffering from light pollution. The trick is to memorize Andromeda's pattern of stars, and then to look slightly to the *side* of where you expect the galaxy to be. This technique – called 'averted vision' – causes the image to fall on the outer region of your retina, which is more sensitive to light than the central region that's evolved to discern fine details. Averted vision is crucial when you want to observe the faintest nebulae or galaxies through a telescope.

OCTOBER'S OBJECT

The star **Algol**, in the constellation **Perseus**, represents the head of the dreadful Gorgon Medusa. In Arabic, its name means 'the demon'. Watch Algol carefully and you'll see why. Every 2 days 21 hours, Algol dims in brightness for several hours – to become as faint as the star lying to its lower right (Gorgonea Tertia).

In 1783, an 18-year-old profoundly deaf amateur astronomer, from York – John Goodricke – discovered Algol's regular changes, and proposed that the star was orbited by a large dark planet that periodically blocks off some of its light. We now know that Algol does indeed have a dim companion blocking its brilliant light; but it's a fainter star, rather than a planet.

OCTOBER'S PICTURE

The **Andromeda Galaxy** is the closest spiral galaxy to our Milky Way. Andromeda has two large satellite galaxies (M32, top, and NGC 205, bottom) which play gravitational tug-of-war with their bigger neighbour. In 3–5 billion years' time, the Milky Way and Andromeda are set to collide, creating a vast elliptical galaxy – Milkomeda.

OCTOBER'S TOPIC
Weird moons

Our Solar System is home to hundreds of moons circling its planets. Our own Moon is astonishing: compared in size to our world, it's vast. The Moon is a quarter of the size of the Earth, making our cosmic pair a 'double planet'.

Mars has two tiny moons – probably captured from the nearby asteroid belt. Phobos – just 27 kilometres long – is in a very low orbit about the Red Planet. It will probably crash into Mars in around 50 million years' time, creating a crater 300 kilometres across.

Jupiter – the giant of cosmic family – is home to at least 70 moons. The most thrilling of them is Io, which boasts sensational volcanoes. It's the most active world in the Solar System, and it erupts plumes of sulphur dioxide that shoot 300 kilometres into space. The moon is pummelled by Jupiter's huge gravitational pull – which stirs up the moon's interior.

Ringworld Saturn is circled by 62 moons. The most iconic is Titan: bigger than the planet Mercury, it's the only world in the Solar System – apart from the Earth – to have an atmosphere of nitrogen. Its thick, orange clouds conceal a complex surface. But in 2005, Europe's Huygens space probe penetrated those murky clouds and landed on Titan. It discovered seas of liquid ethane and methane: ideal environments for the development of future life.

In the early evening, the planets are putting on a show in the west. Later – on a dark night – look overhead for the sensational sight of the **Milky Way** arching upwards. It's a stunning inside perspective on the huge Galaxy that is our home in space.

▼ The sky at 10 pm in mid-November, with Moon positions at three-day intervals either side of Full Moon. The star positions are also correct for 11 pm at

NOVEMBER'S CONSTELLATION

In the northern sky hangs a star pattern making the unmistakable shape of a capital 'W'. To the ancients, this constellation represented Queen **Cassiopeia** of Ethiopia, who ruled with her husband King **Cepheus**.

Cassiopeia misguidedly boasted that her daughter **Andromeda** was more beautiful than the sea nymphs. The sea god, Poseidon, was so incensed that he sent a ravaging monster (**Cetus**) to eat the young people of the country. It could only be appeased by the sacrifice of Andromeda – but she was rescued by the hero **Perseus**. Cassiopeia, Cepheus, Andromeda, Perseus and Cetus are now all immortalized in the heavens.

The Chinese saw Cassiopeia as three star groups, including a chariot and a mountain path. Unusually, the central star in Cassiopeia is universally known today by its Chinese name – **Tsih**, 'the whip'. This star is unstable in brightness. Some 55,000 times brighter than the Sun, it spins around at breakneck pace, flinging out streams of gas.

Cassiopeia has seen two more extreme variable stars – supernovae, where an entire star has blown apart. One was observed by Danish astronomer Tycho Brahe in 1572. The other exploded around 1660 as a surprisingly dim supernova, but its expanding gases form the most prominent radio source in the sky, Cassiopeia A.

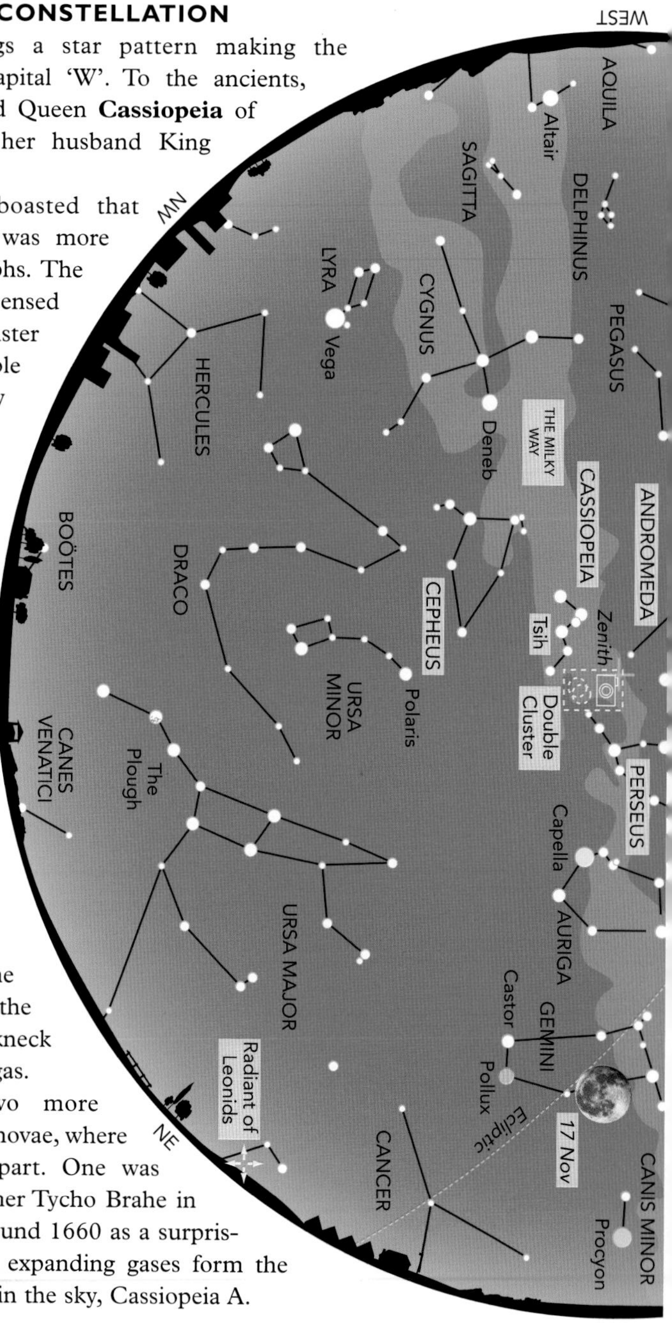

the beginning of November, and 9 pm at the end of the month. The planets move slightly relative to the stars during the month.

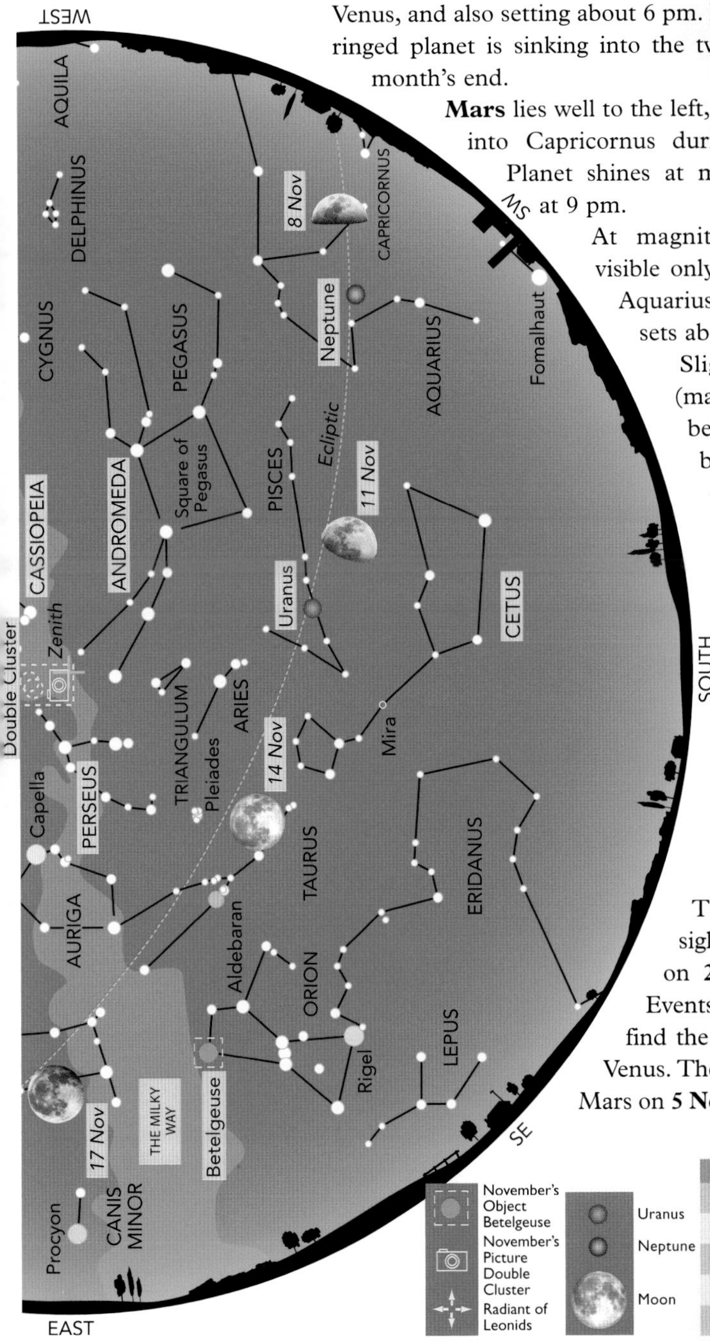

PLANETS ON VIEW

The Evening Star is brilliant in the south-west after sunset: **Venus** is blazing at magnitude −4.0 and sets around 6 pm. It speeds to the left, rising higher, as the month progresses.

At the start of November you'll find Saturn – almost a hundred times fainter at magnitude +0.6 – to the right of Venus, and also setting about 6 pm. Located in Ophiuchus, the ringed planet is sinking into the twilight, and disappears by month's end.

Mars lies well to the left, travelling from Sagittarius into Capricornus during November. The Red Planet shines at magnitude +0.5, and sets at 9 pm.

At magnitude +7.9, **Neptune** is visible only with optical aid: lying in Aquarius, the most remote planet sets about midnight.

Slightly brighter **Uranus** (magnitude +5.7) follows on behind, in Pisces, and drops below the horizon around 4 am.

Early birds can enjoy the sight of **Jupiter**, rising about 3.30 am in the east. Magnificent at magnitude −1.6, the giant planet is currently livening up the constellation of Virgo.

Mercury is lost in the Sun's glare this month.

MOON

The Moon makes a fine sight with Venus and Saturn on **2 November** (see Special Events). On **3 November**, you'll find the Moon well above brilliant Venus. The Moon lies to the right of Mars on **5 November**, and to the upper

MOON		
Date	Time	Phase
7	7.51 pm	First Quarter
14	1.52 pm	Full Moon
21	8.33 am	Last Quarter
29	12.18 pm	New Moon

left of the Red Planet on **6 November**. On the night of **20/21 November**, the Last Quarter Moon moves by Regulus. The morning of **25 November** sees the crescent Moon right next to giant planet Jupiter.

SPECIAL EVENTS

After sunset on **2 November**, look low to the south-west for a pretty line-up of dazzling Venus (left), Saturn and the crescent Moon (right).

The night of **16/17 November** sees the maximum of the **Leonid** meteor shower. This year, the Moon will spoil our view of all but the brightest shooting stars.

NOVEMBER'S OBJECT

Known to generations of school kids as 'Beetlejuice', **Betelgeuse** is one of the biggest stars known. If placed in the Solar System, it would swamp the planets all the way out to the asteroid belt – possibly as far as Jupiter.

Almost 1000 times wider than the Sun, Betelgeuse is a serious red giant – a star close to the end of its life. It's one of just a few stars to be imaged as a visible disc from Earth. Betelgeuse has suffered middle-age spread as frenetic nuclear reactions in the star's core have forced its outer layers to swell and cool. The star also fluctuates slightly in brightness as it tries to get a grip on its billowing gases.

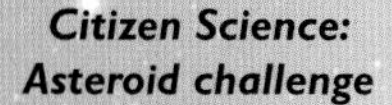

Citizen Science: Asteroid challenge

Asteroids are among the most mysterious bodies in our Solar System. Most of the millions of these irregular chunks of rock – measuring up to several hundred kilometres across – live in the asteroid belt, between Mars and Jupiter. But several thousand cross Earth's orbit, posing a threat to our planet. They may be the primordial material of our Solar System: the debris that accumulated to create the planets. Because they're so small, asteroids are difficult to observe. Their shapes, composition and spin periods are very uncertain. Join Asteroids@home to help astronomers get to grips with these enigmatic bodies – which may give clues to our own origins. **https://asteroidsathome.net/boinc/**

◄ Peter Shah captured this delicate sprinkling of stars making up the Double Cluster in Perseus. NGC 869 lies to the right, and NGC 884 is on the left. Peter took the photo through a 200 mm Newtonian telescope with a CCD camera. He made three separate 10-minute exposures through red, green and blue filters.

Viewing tip
With Christmas coming up, you may be thinking of buying a telescope as a present for a budding stargazer. Beware! Unscrupulous websites and mail-order catalogues selling 'gadgets' often advertise small telescopes that boast huge magnifications. This is known as 'empty magnification' – blowing up an image that the lens or mirror simply doesn't have the ability to get to grips with, so all you see is a bigger blur. As a rule of thumb, the maximum magnification a telescope can actually provide is twice the diameter of the lens or mirror in millimetres. So if you see an advertisement for a 75 mm telescope, beware of any claims for a magnification greater than 150 times.

The origin of the star's name is a mystery. It comes from Arabic, and literally could mean 'the armpit of the sacred one'! But scholars now think the initial 'B' should really be a 'Y', and Betelgeuse is – more boringly – 'the giant's hand'.

Whatever it means, Betelgeuse will exit the Universe in a spectacular supernova explosion. As a result of the breakdown of nuclear reactions at its heart, the star will explode – to shine as brightly in our skies as the Moon.

NOVEMBER'S PICTURE

This beautiful **Double Cluster** in Perseus is a glorious sight in binoculars. Medium-sized telescopes reveal that each cluster contains about 300 stars, but this is only the tip of the iceberg: there are probably thousands of stars in residence. Some 7500 light years away, the blue stars in the Double Cluster are very young (in astronomical terms!) – about 12 million years old. Our Sun, by comparison, is a celestial geriatric: it's been around for nearly 5 *billion* years.

NOVEMBER'S TOPIC
Aurorae

Look out for the sky's ultimate light show this year. The aurora borealis ('northern lights') and its southern equivalent – the aurora australis – are best seen near the Earth's northern and southern poles. In Aberdeen, these arcs, curtains and rays of green, red and yellow shifting lights are called the 'Merrie Dancers'. Aberdonian folklore puts them down to reflections of sunlight off the polar ice-cap.

The cause *is* the Sun – but in a more fundamental way. Our local star is riddled with magnetic fields. As the Sun rotates, its magnetic field is wound up, like a ball of elastic bands. The pressure has to give. Roughly every 11 years, our Sun suffers a serious affliction of dark sunspots, where the magnetism breaks out.

The magnetic loops above the sunspots touch, short-circuit and hurl powerful electrically charged particles into space. If Earth is in the firing line, our planet's magnetic field channels them to the poles, where the electrical particles hit the atmosphere and light up the atoms like gas in a neon tube. Oxygen emits a green colour; nitrogen shines red.

To see the swaying celestial display is an awesome experience – and this is a great time to see the phenomenon, as sunspots are near their peak. You can take special tours by plane to the Arctic Circle – fingers crossed for clear viewing.

And if you can't make the North Pole, never fear. Powerful aurorae have been reported from the south of France – and we've seen one in Oxfordshire!

We have a 'Christmas Star'! Not the Star of Bethlehem (if that existed at all – see this month's Topic), but the brilliant planet Venus, riding high in the evening sky. It's just a foretaste of the glories of these December nights. **Orion**, with his hunting dogs **Canis Major** and **Canis Minor**, is dominating the starry heavens, fighting his adversary **Taurus** (the Bull). He's accompanied by **Castor** and **Pollux**, the twin stars of **Gemini**, along with **Sirius** – the brightest star in the sky – and glorious **Capella**, almost overhead.

▼ The sky at 10 pm in mid-December, with Moon positions at three-day intervals either side of Full Moon. The star positions are also correct for 11 pm at

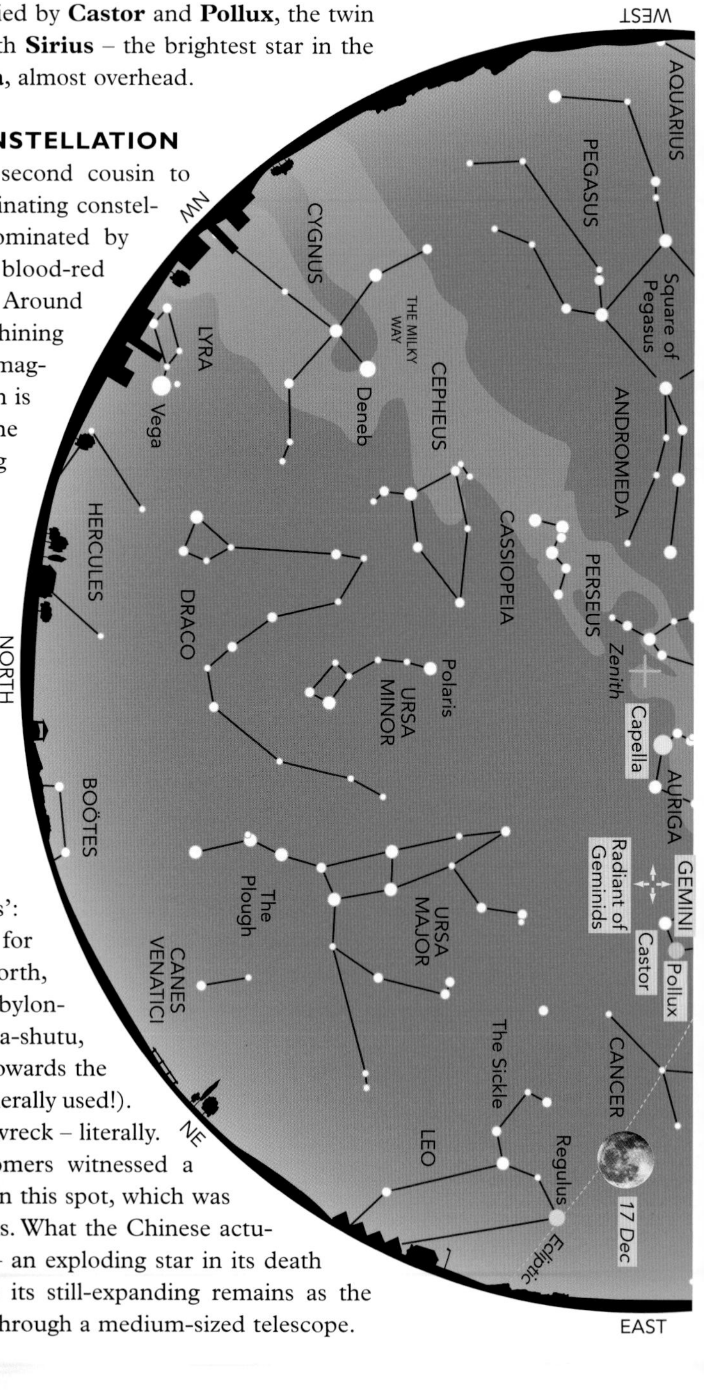

WEST
AQUARIUS
PEGASUS
Square of Pegasus
CYGNUS
THE MILKY WAY
LYRA
Vega
Deneb
CEPHEUS
ANDROMEDA
CASSIOPEIA
HERCULES
DRACO
PERSEUS
Zenith
NORTH
Polaris
URSA MINOR
Capella
AURIGA
BOÖTES
The Plough
URSA MAJOR
Radiant of Geminids
GEMINI
Castor
Pollux
CANES VENATICI
The Sickle
CANCER
LEO
Regulus
17 Dec
Ecliptic
NW
NE
EAST

DECEMBER'S CONSTELLATION

Taurus is very much a second cousin to brilliant **Orion**, but a fascinating constellation nonetheless. It's dominated by **Aldebaran**, the baleful blood-red eye of the celestial bull. Around 65 light years away, and shining with a (slightly variable) magnitude of +0.85, Aldebaran is a red giant star, but not one as extreme as neighbouring **Betelgeuse**. It is slightly more massive than the Sun. The 'head' of the bull is formed by the **Hyades** star cluster. The other famous star cluster in Taurus is the far more glamorous **Pleiades**, whose stars – although further away than the Hyades – are younger and brighter (see this month's Object).

Taurus has two 'horns': the star **El Nath** (Arabic for 'the butting one') to the north, and **zeta Tauri** (whose Babylonian name Shurnarkabti-sha-shutu, meaning 'star in the bull towards the south', is thankfully not generally used!). Above this star is a stellar wreck – literally. In 1054, Chinese astronomers witnessed a brilliant 'new star' appear in this spot, which was visible in daytime for weeks. What the Chinese actually saw was a supernova – an exploding star in its death throes. And today, we see its still-expanding remains as the **Crab Nebula**. It's visible through a medium-sized telescope.

the beginning of December, and 9 pm at the end of the month. The planets move slightly relative to the stars during the month.

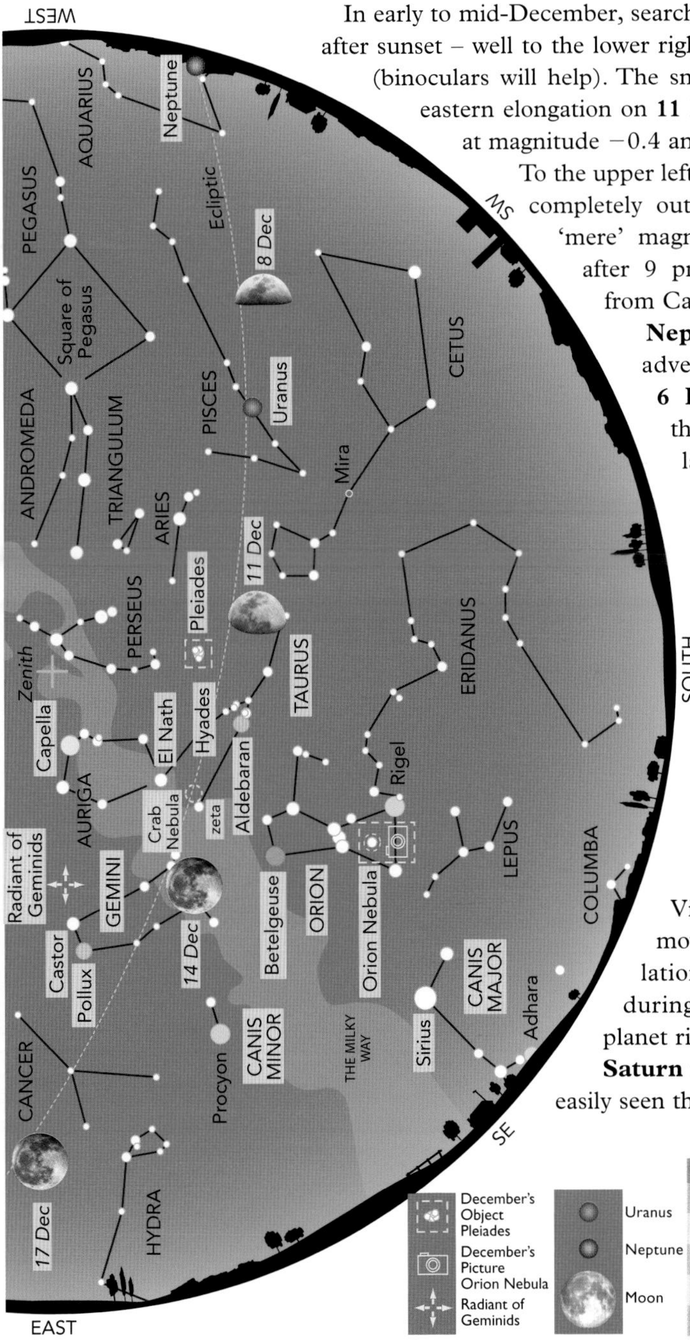

PLANETS ON VIEW

Venus is absolutely stunning in the evening sky. At a dazzling magnitude −4.1, it's putting even the brilliant winter stars to shame. By the end of December, Venus is setting as late as 8 pm and visible in a totally dark sky; away from streetlights, check if you can see shadows cast by the Evening Star.

In early to mid-December, search along the western horizon after sunset – well to the lower right of Venus – for **Mercury** (binoculars will help). The smallest planet is at greatest eastern elongation on **11 December,** when it shines at magnitude −0.4 and sets at 5 pm.

To the upper left of Venus you'll find **Mars,** completely outclassed in brightness at a 'mere' magnitude +0.7. Setting soon after 9 pm, the Red Planet moves from Capricornus into Aquarius.

Neptune is having some adventures this month: on **6 December** it's hidden by the Moon; and on the very last night of 2016, Mars gets up close and personal (see Special Events). You'll need good binoculars or a small telescope to see the distant planet, at magnitude +7.9. Neptune lies in Aquarius, and sets around 10 pm.

Uranus is hanging about in Pisces, at magnitude +5.8, and setting about 2 am.

Mighty **Jupiter** blazes in Virgo at magnitude −1.7, moving towards the constellation's brightest star, Spica, during December. The giant planet rises around 2 am.

Saturn is too close to the Sun to be easily seen this month.

MOON		
Date	Time	Phase
7	9.03 am	First Quarter
14	0.05 am	Full Moon
21	1.55 am	Last Quarter
29	6.53 am	New Moon

MOON

The crescent Moon lies to the right of Venus on **2 December**; the two form a beautiful pair on **3 December** (see Special Events). On **4 December**, the Moon lies between Venus and Mars (to the upper left); and on **5 December**, Mars and the Moon are close. The Moon occults distant Neptune on **6 December** (see Special Events). We're treated to a spectacular occultation of the Hyades and Aldebaran on the night of **12/13 December** (see Special Events). Regulus lies near the Moon on **18 December**. The early hours of **22 December** see the Moon over Jupiter, with Spica below the giant planet; on the morning of **23 December**, the crescent Moon forms a striking triangle with Jupiter (top right) and Spica (lower right).

SPECIAL EVENTS

After sunset on **3 December**, look out for the striking sight of the crescent Moon hanging directly above brilliant Venus.

Turn your binoculars or a telescope towards the Moon on **6 December**: the bluish 'star' to the upper left is Neptune. The Moon is poised to move in front of the planet, in a rare occultation of Neptune that you'll observe sometime between 10.25 pm and 10.40 pm, depending on your location in the UK.

On the night of **12/13 December**, the Moon occults several stars in the Hyades cluster. The event ends with an occultation of Aldebaran (around 5.20 am): the red giant star disappears for about half an hour.

Light pollution from the Full Moon will spoil our enjoyment of the annual **Geminid** meteor shower, on **13/14 December**. But this display (caused by rocky particles from asteroid Phaethon) is noted for bright events, so watch out for the occasional brilliant shooting star!

The Winter Solstice occurs at 10.44 am on **21 December**. As a result of the tilt of Earth's axis, the Sun reaches its lowest point in the heavens as seen from the northern hemisphere: we get the shortest days, and the longest nights.

Find Mars in the west on the evening of **31 December**, and home in with binoculars or a telescope. The faint bluish object just above the Red Planet (less than 20 arc minutes away) is Neptune.

▼ From light-polluted Newport in South Wales, Nick Hart captured a beautiful image of M42, the Orion Nebula. He took the photo with a 200 mm reflector, making separate exposures through red, green, blue, H-alpha and O III filters.

Viewing tip
Venus is a real treat this month, blazing in the evening sky. If you have a small telescope, though, don't wait for the sky to get really dark. Seen against a black sky, the cloud-wreathed planet is so brilliant that it's difficult to make out anything on its disc. It's best to view the planet after the Sun has set, as soon as Venus is visible in the twilight glow: you can then see the planet's disc more clearly against a pale blue sky. Plus, the Evening Star will be higher in the sky and less blurred by turbulent air currents in our atmosphere.

DECEMBER'S OBJECT

The **Pleiades** star cluster is one of the most familiar sky-sights. Though it's well known as the Seven Sisters, skywatchers typically see any number of stars but seven! Most people can pick out the six brightest stars, while very keen-sighted observers can discern up to 11 members. These are just the most luminous in a group of at least 1000 stars. The brightest stars are hot and blue, and all the stars are young – around 100 million years old (about 2% of the Sun's age). They were born together, and have yet to go their separate ways: it will probably take around 250 million years for them all to fly the cosmic nest. The Pleiades are a beautiful sight; and they're magnificent in a long-exposure image (see January's Picture).

DECEMBER'S PICTURE

Grab your chance to spot the iconic **Orion Nebula**. It's visible to the unaided eye below Orion's 'belt', but a telescope shows the huge gas cloud (see January's Object) at its best. Young stars light up the nebulae: the dark zones between them consist of cosmic dust poised to collapse to create new stars. The brightest of these youngsters are the four stars of the Trapezium cluster, easily visible through a small telescope.

DECEMBER'S TOPIC
The Christmas Star

Brilliant Venus, high in the sky this festive month, will focus attention as to the nature of the Christmas Star. What was the object that drew the Magi from the East to Bethlehem?

Chinese astronomers were assiduously recording the sky at that time, and they reported nothing as dazzling as the 'Star' that appears on our Christmas cards.

Perhaps the Magi were inspired by an unusual planetary tango between Jupiter and Saturn in 7 BC: Jupiter was the king of the planets and Saturn the planet of the Jews.

But planetary conjunctions aren't that rare: this year, we've already had Venus skimming closely past Saturn in January, and Jupiter in August. And they don't match the account in St Matthew's gospel – especially of a strange star that 'went before them, till it came and stood over where the young child was'.

We've spent years pondering the Star of Bethlehem. It's not mentioned in any of the other gospels; and Matthew must have been convinced that a once-in-the-Universe event, such as the birth of the Messiah, demanded an amazing celestial portent. So we've concluded there's only one answer to the mystery of the Christmas Star: St Matthew made it up!

There's always something to see in our Solar System, from planets to meteors or the Moon. These objects are very close to us – in astronomical terms – so their positions, shapes and sizes appear to change constantly. It is important to know when, where and how to look if you are to enjoy exploring Earth's neighbourhood. Here we give the best dates in 2016 for observing the planets and meteors (weather permitting!), and explain some of the concepts that will help you to get the most out of your observing.

Magnitudes

Astronomers measure the brightness of stars, planets and other celestial objects using a scale of *magnitudes*. Somewhat confusingly, fainter objects have higher magnitudes, while brighter objects have lower magnitudes; the most brilliant stars have negative magnitudes! Naked-eye stars range from magnitude −1.5 for the brightest star, Sirius, to +6.5 for the faintest stars you can see on a really dark night.

As a guide, here are the magnitudes of selected objects:

Sun	−26.7
Full Moon	−12.5
Venus (at its brightest)	−4.7
Sirius	−1.5
Betelgeuse	+0.4
Polaris (Pole Star)	+2.0
Faintest star visible to the naked eye	+6.5
Faintest star visible to the Hubble Space Telescope	+31

THE INFERIOR PLANETS

A planet with an orbit that lies closer to the Sun than the orbit of Earth is known as *inferior*. Mercury and Venus are the inferior planets. They show a full range of phases (like the Moon) from the thinnest crescents to full, depending on their position in relation to the Earth and the Sun. The diagram below shows the various positions of the inferior planets. They are invisible when at *conjunction*, when they are either behind the Sun, or between the Earth and the Sun, and lost in the latter's glare.

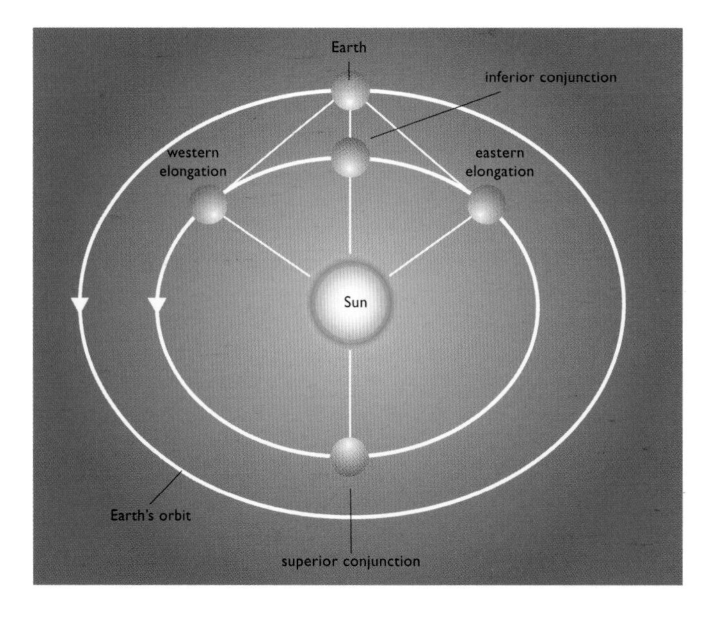

◀ *At eastern or western elongation, an inferior planet is at its maximum angular distance from the Sun. Conjunction occurs at two stages in the planet's orbit. Under certain circumstances, an inferior planet can transit across the Sun's disc at inferior conjunction.*

Mercury

Shy little Mercury puts on a star turn this year, when it transits across the face of the Sun on **9 May** (see May's Topic). In the evening sky, you'll see Mercury best in mid-April; we also have reasonable views at the beginning of January and in December. Early birds can spot Mercury before dawn in January–February, with its best morning appearance coming in September–October. (The apparitions in June and August are low in bright twilight sky.)

Maximum elongations of Mercury in 2016

Date	Separation
7 February	26° west
18 April	20° east
5 June	24° west
16 August	27° east
28 September	18° west
11 December	21° east

Venus

Venus swings behind the Sun on **6 June**, and it's relatively inconspicuous for most of the year: the planet has no elongations in 2016. Appearing as the Morning Star at the start of 2016, Venus lurks near the horizon until September, when it starts surging upwards to emerge as a brilliant Evening Star in December.

THE SUPERIOR PLANETS

The superior planets are those with orbits that lie beyond that of the Earth. They are Mars, Jupiter, Saturn, Uranus and Neptune. The best time to observe a superior planet is when the Earth lies between it and the Sun. At this point in a planet's orbit, it is said to be at *opposition*.

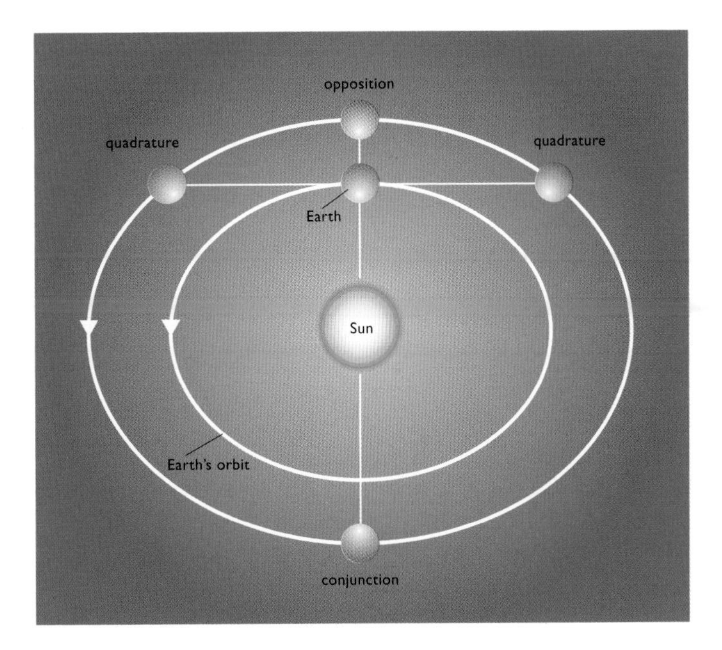

► *Superior planets are invisible at conjunction. At quadrature the planet is at right angles to the Sun as viewed from Earth. Opposition is the best time to observe a superior planet.*

Progress of Mars through the constellations

Early January	Virgo
Mid-Jan to mid-March	Libra
Mid-March to May	Scorpius
June to July	Libra
August	Scorpius
September	Ophiuchus
October	Sagittarius
November to mid-Dec	Capricornus
Late December	Aquarius

Mars

The Red Planet is in spectacular form in 2016. Reaching opposition on **22 May**, eight days later (thanks to its elliptical orbit) Mars is at its closest to the Earth in ten years, and outshines everything else in the night sky bar the Moon.

Jupiter

The giant planet blazes in the evening sky at the start of 2016. It inhabits Leo from January, through its opposition on **8 March**, until it disappears into the evening twilight glow at the end of August (after an amazingly close brush with Venus on **27 August**). Jupiter reappears in the evening sky in October, in the constellation Virgo.

Saturn

The ringed planet lies low in Ophiuchus throughout the year. Saturn is prominent in the morning sky at the start of 2016; is visible all night when it reaches opposition on **3 June**; and remains a feature of the evening sky until November.

Uranus

Just perceptible to the naked eye, Uranus is visible in the evening sky from January to March; and then from July to December. It swims among the stars of Pisces all year, and is at opposition on **15 October**.

Neptune

Lying in Aquarius all year, the most distant planet is at opposition on **2 September**. Neptune can be seen – though only through binoculars or a telescope – in January; and then from April to the end of the year.

Astronomical distances

For objects in the Solar System, such as the planets, we can give their distances from the Earth in kilometres. But the distances are just too huge once we reach out to the stars. Even the nearest star (Proxima Centauri) lies 25 million million kilometres away.

So astronomers use a larger unit – the *light year.* This is the distance that light travels in one year, and it equals 9.46 million million kilometres.

Here are the distances to some familiar astronomical objects, in light years:

Proxima Centauri	4.2
Betelgeuse	640
Centre of the Milky Way	27,000
Andromeda Galaxy	2.5 million
Most distant galaxies seen by the Hubble Space Telescope	13 billion

SOLAR AND LUNAR ECLIPSES

Solar Eclipses

A total solar eclipse on **9 March** is visible from a narrow region stretching from Sumatra, through Borneo and Sulawesi, to the Pacific Ocean north of Hawaii. People in Indochina, northern Australasia and the north-west Pacific will witness a partial eclipse. The track of an annular eclipse on **1 September** passes across Africa from Gabon to Madagascar. The rest of Africa and the Indian Ocean experience a partial solar eclipse.

Lunar Eclipses

There are no total or partial eclipses of the Moon this year. Two penumbral eclipses (**23 March** and **16 September**) – when the Moon is slightly shaded from the Sun's light – will not be noticeable from Britain.

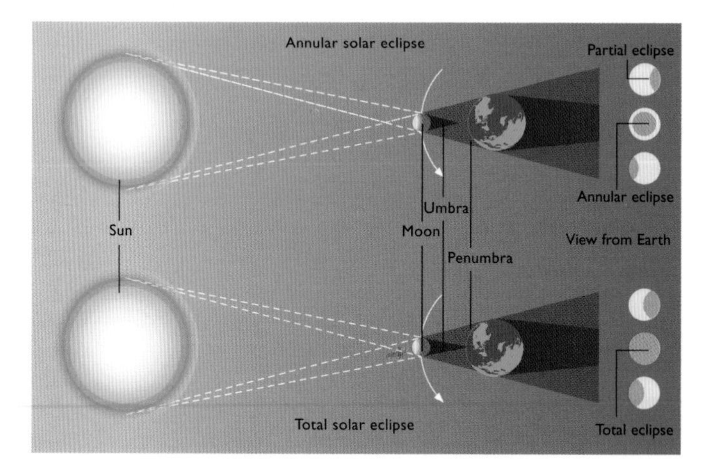

◀ *Where the dark central part (the umbra) of the Moon's shadow reaches the Earth, we see a total eclipse. People located within the penumbra see a partial eclipse. If the umbral shadow does not reach the Earth, we see an annular eclipse. This type of eclipse occurs when the Moon is at a distant point in its orbit and is not quite large enough to cover the whole of the Sun's disc.*

METEOR SHOWERS

Dates of maximum for selected meteor showers	
Meteor shower	Date of maximum
Quadrantids	3/4 January
Lyrids	21/22 April
Eta Aquarids	5/6 May
Perseids	11/12 August
Orionids	20/21 October
Leonids	16/17 November
Geminids	13/14 December

Shooting stars – or *meteors* – are tiny particles of interplanetary dust, known as *meteoroids*, burning up in the Earth's atmosphere. At certain times of year, the Earth passes through a stream of these meteoroids (usually debris left behind by a comet) and we see a *meteor shower*. The point in the sky from which the meteors appear to emanate is known as the *radiant*. Most showers are known by the constellation in which the radiant is situated.

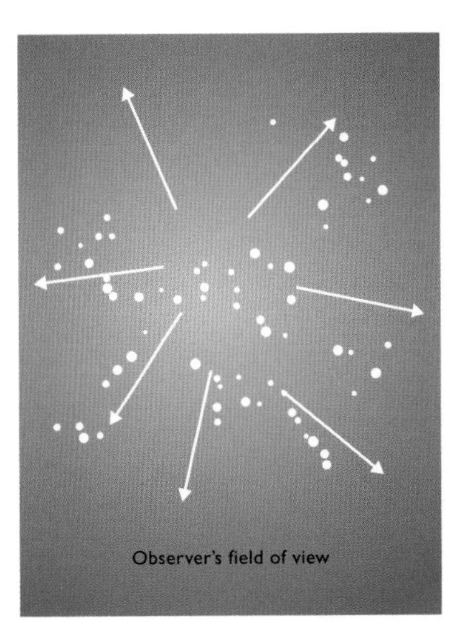

► *Meteors from a common source, occurring during a shower, enter the atmosphere along parallel trajectories. As a result of perspective, however, they appear to diverge from a single point in the sky – the radiant.*

When watching meteors for a co-ordinated meteor programme, observers generally note the time, seeing conditions, cloud cover, their own location, the time and brightness of each meteor, and whether it was from the main meteor stream. It is also worth noting details of persistent afterglows (trains) and fireballs, and making counts of how many meteors appear in a given period.

Angular separations

Astronomers measure the distance between objects, as we see them in the sky, by the angle between the objects in degrees (symbol °). From the horizon to the point above your head is 90 degrees. All around the horizon is 360 degrees.

You can use your hand, held at arm's length, as a rough guide to angular distances, as follows:

Width of index finger 1°
Width of clenched hand 10°
Thumb to little finger on outspread hand 20°

For smaller distances, astronomers divide the degree into 60 arc minutes (symbol ′), and the arc minute into 60 arc seconds (symbol ″).

COMETS

Comets are small bodies in orbit about the Sun. Consisting of frozen gases and dust, they are often known as 'dirty snowballs'. When their orbits bring them close to the Sun, the ices evaporate and dramatic tails of gas and dust can sometimes be seen.

A number of comets move round the Sun in fairly small, elliptical orbits in periods of a few years; others have much longer periods. Most really brilliant comets have orbital periods of several thousands or even millions of years. The exception is Comet Halley, a bright comet with a period of about 76 years. It was last seen with the naked eye in 1986.

Binoculars and wide-field telescopes provide the best views of comet tails. Larger telescopes with a high magnification are necessary to observe fine detail in the gaseous head (*coma*). Most comets are discovered with professional instruments, but a few are still found by experienced amateur astronomers.

In January, Comet Catalina should be faintly visible to the naked eye; you can follow it through to mid-February with binoculars. Though astronomers don't currently know of any other comets that will reach naked-eye brightness in 2016, a brilliant new comet could always put in a surprise appearance.

Deep-sky objects are 'fuzzy patches' that lie outside the Solar System. They include star clusters, nebulae and galaxies. To observe the majority of deep-sky objects you will need binoculars or a telescope, but there are also some beautiful naked-eye objects, notably the Pleiades and the Orion Nebula.

The faintest object that an instrument can see is its *limiting magnitude*. The table gives a rough guide, for good seeing conditions, for a variety of small- to medium-sized telescopes.

We have provided a selection of recommended deep-sky targets, together with their magnitudes. Some are described in more detail in our monthly 'Object' features. Look on the appropriate month's map to find which constellations are on view, and then choose your objects using the list below. We have provided celestial coordinates for readers with detailed star maps or Go To telescopes. The suggested times of year for viewing are when the constellation is highest in the sky in the late evening.

Limiting magnitude for small to medium telescopes

Aperture (mm)	Limiting magnitude
50	+11.2
60	+11.6
70	+11.9
80	+12.2
100	+12.7
125	+13.2
150	+13.6

RECOMMENDED DEEP-SKY OBJECTS

Andromeda – autumn and early winter	
M31 (NGC 224) Andromeda Galaxy	3rd-magnitude spiral galaxy *RA 00h 42.7m Dec +41° 16'*
M32 (NGC 221)	8th-magnitude elliptical galaxy, a companion to M31 *RA 00h 42.7m Dec +40° 52'*
M110 (NGC 205)	8th-magnitude elliptical galaxy *RA 00h 40.4m Dec +41° 41'*
NGC 7662 Blue Snowball	8th-magnitude planetary nebula *RA 23h 25.9m Dec +42° 33'*
Aquarius – late autumn and early winter	
M2 (NGC 7089)	6th-magnitude globular cluster *RA 21h 33.5m Dec –00° 49'*
M72 (NGC 6981)	9th-magnitude globular cluster *RA 20h 53.5m Dec –12° 32'*
NGC 7293 Helix Nebula	7th-magnitude planetary nebula *RA 22h 29.6m Dec –20° 48'*
NGC 7009 Saturn Nebula	8th-magnitude planetary nebula *RA 21h 04.2m Dec –11° 22'*
Aries – early winter	
NGC 772	10th-magnitude spiral galaxy *RA 01h 59.3m Dec +19° 01'*
Auriga – winter	
M36 (NGC 1960)	6th-magnitude open cluster *RA 05h 36.1m Dec +34° 08'*
M37 (NGC 2099)	6th-magnitude open cluster *RA 05h 52.4m Dec +32° 33'*
M38 (NGC 1912)	6th-magnitude open cluster *RA 05h 28.7m Dec +35° 50'*
Cancer – late winter to early spring	
M44 (NGC 2632) Praesepe or Beehive	3rd-magnitude open cluster *RA 08h 40.1m Dec +19° 59'*
M67 (NGC 2682)	7th-magnitude open cluster *RA 08h 50.4m Dec +11° 49'*
Canes Venatici – visible all year	
M3 (NGC 5272)	6th-magnitude globular cluster *RA 13h 42.2m Dec +28° 23'*
M51 (NGC 5194/5) Whirlpool Galaxy	8th-magnitude spiral galaxy *RA 13h 29.9m Dec +47° 12'*
M63 (NGC 5055)	9th-magnitude spiral galaxy *RA 13h 15.8m Dec +42° 02'*
M94 (NGC 4736)	8th-magnitude spiral galaxy *RA 12h 50.9m Dec +41° 07'*
M106 (NGC4258)	8th-magnitude spiral galaxy *RA 12h 19.0m Dec +47° 18'*
Canis Major – late winter	
M41 (NGC 2287)	4th-magnitude open cluster *RA 06h 47.0m Dec –20° 44'*
Capricornus – late summer and early autumn	
M30 (NGC 7099)	7th-magnitude globular cluster *RA 21h 40.4m Dec –23° 11'*
Cassiopeia – visible all year	
M52 (NGC 7654)	6th-magnitude open cluster *RA 23h 24.2m Dec +61° 35'*
M103 (NGC 581)	7th-magnitude open cluster *RA 01h 33.2m Dec +60° 42'*
NGC 225	7th-magnitude open cluster *RA 00h 43.4m Dec +61 47'*
NGC 457	6th-magnitude open cluster *RA 01h 19.1m Dec +58° 20'*
NGC 663	Good binocular open cluster *RA 01h 46.0m Dec +61° 15'*
Cepheus – visible all year	
Delta Cephei	Variable star, varying between +3.5 and +4.4 with a period of 5.37 days. It has a magnitude +6.3 companion and they make an attractive pair for small telescopes or binoculars.
Cetus – late autumn	
Mira (omicron Ceti)	Irregular variable star with a period of roughly 330 days and a range between +2.0 and +10.1.
M77 (NGC 1068)	9th-magnitude spiral galaxy *RA 02h 42.7m Dec –00° 01'*

Coma Berenices – spring

M53 (NGC 5024)	8th-magnitude globular cluster *RA 13h 12.9m Dec +18° 10′*
M64 (NGC 4286) Black Eye Galaxy	8th-magnitude spiral galaxy with a prominent dust lane that is visible in larger telescopes. *RA 12h 56.7m Dec +21° 41′*
M85 (NGC 4382)	9th-magnitude elliptical galaxy *RA 12h 25.4m Dec +18° 11′*
M88 (NGC 4501)	10th-magnitude spiral galaxy *RA 12h 32.0m Dec.+14° 25′*
M91 (NGC 4548)	10th-magnitude spiral galaxy *RA 12h 35.4m Dec +14° 30′*
M98 (NGC 4192)	10th-magnitude spiral galaxy *RA 12h 13.8m Dec +14° 54′*
M99 (NGC 4254)	10th-magnitude spiral galaxy *RA 12h 18.8m Dec +14° 25′*
M100 (NGC 4321)	9th-magnitude spiral galaxy *RA 12h 22.9m Dec +15° 49′*
NGC 4565	10th-magnitude spiral galaxy *RA 12h 36.3m Dec +25° 59′*

Cygnus – late summer and autumn

Cygnus Rift	Dark cloud just south of Deneb that appears to split the Milky Way in two.
NGC 7000 North America Nebula	A bright nebula against the background of the Milky Way, visible with binoculars under dark skies. *RA 20h 58.8m Dec +44° 20′*
NGC 6992 Veil Nebula (part)	Supernova remnant, visible with binoculars under dark skies. *RA 20h 56.8m Dec +31 28′*
M29 (NGC 6913)	7th-magnitude open cluster *RA 20h 23.9m Dec +36° 32′*
M39 (NGC 7092)	Large 5th-magnitude open cluster *RA 21h 32.2m Dec +48° 26′*
NGC 6826 Blinking Planetary	9th-magnitude planetary nebula *RA 19 44.8m Dec +50° 31′*

Delphinus – late summer

NGC 6934	9th-magnitude globular cluster *RA 20h 34.2m Dec +07° 24′*

Draco – midsummer

NGC 6543	9th-magnitude planetary nebula *RA 17h 58.6m Dec +66° 38′*

Gemini – winter

M35 (NGC 2168)	5th-magnitude open cluster *RA 06h 08.9m Dec +24° 20′*
NGC 2392 Eskimo Nebula	8–10th-magnitude planetary nebula *RA 07h 29.2m Dec +20° 55′*

Hercules – early summer

M13 (NGC 6205)	6th-magnitude globular cluster *RA 16h 41.7m Dec +36° 28′*
M92 (NGC 6341)	6th-magnitude globular cluster *RA 17h 17.1m Dec +43° 08′*
NGC 6210	9th-magnitude planetary nebula *RA 16h 44.5m Dec +23 49′*

Hydra – early spring

M48 (NGC 2548)	6th-magnitude open cluster *RA 08h 13.8m Dec –05° 48′*
M68 (NGC 4590)	8th-magnitude globular cluster *RA 12h 39.5m Dec –26° 45′*
M83 (NGC 5236)	8th-magnitude spiral galaxy *RA 13h 37.0m Dec –29° 52′*
NGC 3242 Ghost of Jupiter	9th-magnitude planetary nebula *RA 10h 24.8m Dec –18° 38′*

Leo – spring

M65 (NGC 3623)	9th-magnitude spiral galaxy *RA 11h 18.9m Dec +13° 05′*
M66 (NGC 3627)	9th-magnitude spiral galaxy *RA 11h 20.2m Dec +12° 59′*
M95 (NGC 3351)	10th-magnitude spiral galaxy *RA 10h 44.0m Dec +11° 42′*
M96 (NGC 3368)	9th-magnitude spiral galaxy *RA 10h 46.8m Dec +11° 49′*
M105 (NGC 3379)	9th-magnitude elliptical galaxy *RA 10h 47.8m Dec +12° 35′*

Lepus – winter

M79 (NGC 1904)	8th-magnitude globular cluster *RA 05h 24.5m Dec –24° 33′*

Lyra – spring

M56 (NGC 6779)	8th-magnitude globular cluster *RA 19h 16.6m Dec +30° 11′*
M57 (NGC 6720) Ring Nebula	9th-magnitude planetary nebula *RA 18h 53.6m Dec +33° 02′*

Monoceros – winter

M50 (NGC 2323)	6th-magnitude open cluster *RA 07h 03.2m Dec –08° 20′*
NGC 2244	Open cluster surrounded by the faint Rosette Nebula, NGC 2237. Visible in binoculars. *RA 06h 32.4m Dec +04° 52′*

Ophiuchus – summer

M9 (NGC 6333)	8th-magnitude globular cluster *RA 17h 19.2m Dec –18° 31′*
M10 (NGC 6254)	7th-magnitude globular cluster *RA 16h 57.1m Dec –04° 06′*
M12 (NCG 6218)	7th-magnitude globular cluster *RA 16h 47.2m Dec –01° 57′*
M14 (NGC 6402)	8th-magnitude globular cluster *RA 17h 37.6m Dec –03° 15′*
M19 (NGC 6273)	7th-magnitude globular cluster *RA 17h 02.6m Dec –26° 16′*
M62 (NGC 6266)	7th-magnitude globular cluster *RA 17h 01.2m Dec –30° 07′*
M107 (NGC 6171)	8th-magnitude globular cluster *RA 16h 32.5m Dec –13° 03′*

Orion – winter

M42 (NGC 1976) Orion Nebula	4th-magnitude nebula *RA 05h 35.4m Dec –05° 27′*
M43 (NGC 1982)	5th-magnitude nebula *RA 05h 35.6m Dec –05° 16′*
M78 (NGC 2068)	8th-magnitude nebula *RA 05h 46.7m Dec +00° 03′*

Pegasus – autumn

M15 (NGC 7078)	6th-magnitude globular cluster *RA 21h 30.0m Dec +12° 10′*

Perseus – autumn to winter

M34 (NGC 1039)	5th-magnitude open cluster *RA 02h 42.0m Dec +42° 47′*
M76 (NGC 650/1) Little Dumbbell	11th-magnitude planetary nebula *RA 01h 42.4m Dec +51° 34′*

NGC 869/884 Double Cluster	Pair of open star clusters *RA 02h 19.0m Dec +57° 09′* *RA 02h 22.4m Dec +57° 07′*
Pisces – autumn	
M74 (NGC 628)	9th-magnitude spiral galaxy *RA 01h 36.7m Dec +15° 47′*
Puppis – late winter	
M46 (NGC 2437)	6th-magnitude open cluster *RA 07h 41.8m Dec –14° 49′*
M47 (NGC 2422)	4th-magnitude open cluster *RA 07h 36.6m Dec –14° 30′*
M93 (NGC 2447)	6th-magnitude open cluster *RA 07h 44.6m Dec –23° 52′*
Sagitta – late summer	
M71 (NGC 6838)	8th-magnitude globular cluster *RA 19h 53.8m Dec +18° 47′*
Sagittarius – summer	
M8 (NGC 6523) Lagoon Nebula	6th-magnitude nebula *RA 18h 03.8m Dec –24° 23′*
M17 (NGC 6618) Omega Nebula	6th-magnitude nebula *RA 18h 20.8m Dec –16° 11′*
M18 (NGC 6613)	7th-magnitude open cluster *RA 18h 19.9m Dec –17 08′*
M20 (NGC 6514) Trifid Nebula	9th-magnitude nebula *RA 18h 02.3m Dec –23° 02′*
M21 (NGC 6531)	6th-magnitude open cluster *RA 18h 04.6m Dec –22° 30′*
M22 (NGC 6656)	5th-magnitude globular cluster *RA 18h 36.4m Dec –23° 54′*
M23 (NGC 6494)	5th-magnitude open cluster *RA 17h 56.8m Dec –19° 01′*
M24 (NGC 6603)	5th-magnitude open cluster *RA 18h 16.9m Dec –18° 29′*
M25 (IC 4725)	5th-magnitude open cluster *RA 18h 31.6m Dec –19° 15′*
M28 (NGC 6626)	7th-magnitude globular cluster *RA 18h 24.5m Dec –24° 52′*
M54 (NGC 6715)	8th-magnitude globular cluster *RA 18h 55.1m Dec –30° 29′*
M55 (NGC 6809)	7th-magnitude globular cluster *RA 19h 40.0m Dec –30° 58′*
M69 (NGC 6637)	8th-magnitude globular cluster *RA 18h 31.4m Dec –32° 21′*
M70 (NGC 6681)	8th-magnitude globular cluster *RA 18h 43.2m Dec –32° 18′*
M75 (NGC 6864)	9th-magnitude globular cluster *RA 20h 06.1m Dec –21° 55′*
Scorpius (northern part) – midsummer	
M4 (NGC 6121)	6th-magnitude globular cluster *RA 16h 23.6m Dec –26° 32′*
M7 (NGC 6475)	3rd-magnitude open cluster *RA 17h 53.9m Dec –34° 49′*
M80 (NGC 6093)	7th-magnitude globular cluster *RA 16h 17.0m Dec –22° 59′*
Scutum – mid to late summer	
M11 (NGC 6705) Wild Duck Cluster	6th-magnitude open cluster *RA 18h 51.1m Dec –06° 16′*
M26 (NGC 6694)	8th-magnitude open cluster *RA 18h 45.2m Dec –09° 24′*
Serpens – summer	
M5 (NGC 5904)	6th-magnitude globular cluster *RA 15h 18.6m Dec +02° 05′*
M16 (NGC 6611)	6th-magnitude open cluster, surrounded by the Eagle Nebula. *RA 18h 18.8m Dec –13° 47′*
Taurus – winter	
M1 (NGC 1952) Crab Nebula	8th-magnitude supernova remnant *RA 05h 34.5m Dec +22° 00′*
M45 Pleiades	1st-magnitude open cluster, an excellent binocular object. *RA 03h 47.0m Dec +24° 07′*
Triangulum – autumn	
M33 (NGC 598)	6th-magnitude spiral galaxy *RA 01h 33.9m Dec +30° 39′*
Ursa Major – all year	
M81 (NGC 3031)	7th-magnitude spiral galaxy *RA 09h 55.6m Dec +69° 04′*
M82 (NGC 3034)	8th-magnitude starburst galaxy *RA 09h 55.8m Dec +69° 41′*
M97 (NGC 3587) Owl Nebula	12th-magnitude planetary nebula *RA 11h 14.8m Dec +55° 01′*
M101 (NGC 5457)	8th-magnitude spiral galaxy *RA 14h 03.2m Dec +54° 21′*
M108 (NGC 3556)	10th-magnitude spiral galaxy *RA 11h 11.5m Dec +55° 40′*
M109 (NGC 3992)	10th-magnitude spiral galaxy *RA 11h 57.6m Dec +53° 23′*
Virgo – spring	
M49 (NGC 4472)	8th-magnitude elliptical galaxy *RA 12h 29.8m Dec +08° 00′*
M58 (NGC 4579)	10th-magnitude spiral galaxy *RA 12h 37.7m Dec +11° 49′*
M59 (NGC 4621)	10th-magnitude elliptical galaxy *RA 12h 42.0m Dec +11° 39′*
M60 (NGC 4649)	9th-magnitude elliptical galaxy *RA 12h 43.7m Dec +11° 33′*
M61 (NGC 4303)	10th-magnitude spiral galaxy *RA 12h 21.9m Dec +04° 28′*
M84 (NGC 4374)	9th-magnitude elliptical galaxy *RA 12h 25.1m Dec +12° 53′*
M86 (NGC 4406)	9th-magnitude elliptical galaxy *RA 12h 26.2m Dec +12° 57′*
M87 (NGC 4486)	9th-magnitude elliptical galaxy *RA 12h 30.8m Dec +12° 24′*
M89 (NGC 4552)	10th-magnitude elliptical galaxy *RA 12h 35.7m Dec +12° 33′*
M90 (NGC 4569)	9th-magnitude spiral galaxy *RA 12h 36.8m Dec +13° 10′*
M104 (NGC 4594) Sombrero Galaxy	Almost edge-on 8th-magnitude spiral galaxy. *RA 12h 40.0m Dec –11° 37′*
Vulpecula – late summer and autumn	
M27 (NGC 6853) Dumbbell Nebula	8th-magnitude planetary nebula *RA 19h 59.6m Dec +22° 43′*

THE PLANETARY IMAGING REVOLUTION

Until about 2000, even photos of the planets taken at mountain-top observatories using large telescopes were of poor quality compared with the visual view. The standard advice given in books on amateur astronomy was that planetary observing was an area where the visual observer could still play a part, because the eye could catch those moments of atmospheric steadiness which photographs would miss. Earth's unsteady atmosphere meant that fine detail was always smeared out by the comparatively long exposures needed to capture an image on film.

Even the CCD revolution of the 1990s didn't make a great difference. Early CCDs were more sensitive than film, and amateurs began to get improved images, but it was still quite hit-or-miss whether they got a moment of steadiness. What observers call 'good seeing' – that is, a steady atmosphere – was still hard to capture.

But today, amateurs in their back gardens regularly take planetary images which are not only better than the very best from dedicated observatories during the last century, but seem to be better than is theoretically possible with comparatively small instruments. What has made the difference, and how can mere amateurs get better results than theory allows?

It was when amateurs got their hands on webcams around 2000 that everything changed. These gave a low-resolution video stream which could be transmitted over the slow modems of the time so that computer users could see each other. The resolution was either 320 × 240 pixels, or 640 × 480 pixels – poor by pictorial standards, but perfectly acceptable for video. They contained CCD chips, in some cases with as much sensitivity as those being used for serious astronomy. Being mass-produced, they were cheap. Amateurs began to experiment with what they could do when linked to a telescope.

► *This image of Saturn by Damian Peach using a C14 355 mm telescope from Barbados shows the Encke Division in its outer A Ring. This is only some 500 km wide, and should require about a 600 mm telescope to be visible, according to the Dawes' Limit. Photo taken using a ZWO ASI174MM camera.*

The video stream contained both good and bad individual images, and ingenious software was devised to select the best frames and stack them, so reducing the noise level in the individual frames, and some astounding results emerged.

The success of the method is due partly to the improved sensitivity, compared with film, of electronic detectors and partly to the difference between digital processing and visual observing. Those fortunate brief glimpses of fine detail that visual observers could only attempt to capture in a drawing could now be preserved and amplified – a method sometimes referred to as 'lucky imaging'. Fine contrast differences could be enhanced using the software.

▲ *The Sunflower Galaxy, M63, photographed by Ian Papworth using a mono ASI120MM camera and filter wheel through a 150 mm Celestron NexStar 6SE telescope. He combined 230 separate 10-second exposures to overcome the image rotation that would otherwise have resulted from the altazimuth mounting.*

The practical performance limit of a particular aperture, known as Dawes' Limit, was originally devised by using a small telescope visually to separate the components of double stars, and extrapolating this to larger apertures. There are theoretical limits as well, based on the properties of light when it passes through a small aperture. The best known of these, the Rayleigh Criterion, uses a fairly arbitrary basis for what is visible. However, with image processing, not only is it possible to choose moments of good seeing, but the criteria for distinguishing the gap between the stars are different. So what seemed to be a contradiction of the laws of physics is more to do with the choice of criteria for distinguishing fine detail.

In practice, images of Jupiter from amateur back-garden instruments can show detail only 0.1 arc seconds across, even though typical seeing in the UK is about 2 arc seconds and the theoretical resolving power of a typical 355 mm telescope such as a Celestron C14 is only 0.32 arc seconds. The visibility of Encke's Division in Saturn's A ring in images taken with telescopes as small as 250 mm is further proof of the power of modern imaging techniques.

PLANETARY IMAGING FOR ALL

The good news is that it is now relatively easy using quite modest instruments to take images of the Moon and planets that would have amazed amateurs just 20 years ago. Having said that, taking the very best images still requires great effort and attention to detail. But for most people, getting a result that looks good on the wall is what it's about, even if the most dedicated imagers can do better.

So what's needed? A telescope, obviously. This doesn't need to be anything grand or expensive, though the results from instruments below about 125 mm aperture will not be very exciting. It should have a driven mounting, but for planetary work you don't need the sort of precise drive that you'd need for deep-sky imaging. A motor drive that keeps the planet in the field of view for a few minutes at a time is good enough.

And you need a suitable camera. The webcam imaging revolution began with good-quality off-the-shelf webcams costing around £40, such as the legendary Philips ToUcam, but these are no longer available. People can sometimes get impressive results with phone or compact cameras, just pointing them into the telescope, but again these are not ideal. A more useful alternative is a DSLR camera. Here, the problem is that the planet occupies only a tiny part of the field of view, so you need to use pre-processing software to cut down the size of the stream to manageable proportions. It's doable, and it makes use of a camera that you may have already, but doesn't usually give the best results. Most planetary imagers now opt for a purpose-built camera.

THE CAMERA

Over the years, several companies have made cameras which are ideal for the purpose. Point Grey, The Imaging Source and Celestron's Skyris range all have their devotees. But one Chinese company in particular, ZWO, has brought down the cost of good-quality planetary cameras. Not only do their ASI cameras give excellent results on the planets, they can also be used for longer exposures to take deep-sky images, and some even come equipped with wide-angle lenses so they can be used for wide-field imaging out of the box.

A computer is essential for operating these cameras. Software to run them is provided, or can be downloaded, Firecapture being particularly designed for astronomical work. Most are available in colour or mono versions. The latter must be used with individual colour filters, usually mounted in a filter wheel, to take separate red, green and blue sequences of images.

► The ZWO ASI120MC camera comes with a lens that gives a 150° field of view for wide-field imaging. Exposure times up to 600 seconds are possible.

Advanced imagers prefer this method of working, as they can choose the individual filters and can if necessary use wavebands beyond the usual visual range, such as the near infrared and ultraviolet.

But for many, the basic ZWO ASI120MC camera, giving colour results, is all they need, and in tests the improvements gained by using a mono camera with filters are often quite small except under ideal conditions. The recently introduced ASI174MM (mono) and ASI174MC cameras offer USB 3.0 performance for those whose computers are suitably equipped, larger chips and lower noise. The 120 series cameras cost about £225 and the 174 cameras about £550 at 2015 prices.

You'll still need separate image-processing software, such as Registax or Autostakkert 2, but these are freely available. There is a learning curve, as with everything astronomical, but there's plenty of online help.

During a typical observing session you first get your planet central in your field of view, using an eyepiece, and then replace the eyepiece with your camera. Hopefully you'll see an image on your laptop, though you'll probably have to refocus. Start with a fairly long exposure time of about ¼ second and you'll see the disc of light of the out-of-focus planet quite quickly, which you can then focus. Then adjust the exposure time to get good details on the planet. The image will probably be quite small, so a Barlow lens is usually needed to increase the image size and bring out the fine detail.

For deep-sky images a better drive is needed, but by stacking even 15-second images it's possible to achieve worthwhile results. A cooled CCD camera will do better, but at considerably higher cost.

While the outlay is not trivial, as with the basic webcams, the versatility and quality of these cameras makes them very popular with both beginners and advanced imagers alike.

USEFUL WEBSITES

ZWO website, for technical information:
astronomy-imaging-camera.com/

UK suppliers:
ZWO products are available from several specialist UK astronomy suppliers.

Registax stacking software:
www.astronomie.be/registax/

Autostakkert 2 stacking software:
www.autostakkert.com/

◀ A typical imaging set-up using a mono camera, with filter wheel carrying red, green and blue filters, and a Barlow lens to enlarge the tiny planetary disc.